Ecologists and Ethical Judgements

Ecologists and Ethical Judgements

Edited by

N.S. Cooper
Rector
Rivenhall
Witham
Essex, UK

and

R.C.J. Carling
Senior Commissioning Editor
Life Sciences
Chapman & Hall
London, UK

CHAPMAN & HALL
London · Weinheim · New York · Tokyo · Melbourne · Madras

Published by Chapman & Hall, 2–6 Boundary Row, London SE1 8HN

Chapman & Hall, 2–6 Boundary Row, London SE1 8HN, UK

Chapman & Hall GmbH, Pappelallee 3, 69469 Weinheim, Germany

Chapman & Hall USA, 115 Fifth Avenue, New York NY 10003, USA

Chapman & Hall Japan, ITP-Japan, Kyowa Building, 3F, 2-2-1, Hirakawacho, Chiyoda-ku, Tokyo 102, Japan

Chapman & Hall Australia, 102 Dodds Street, South Melbourne, Victoria 3205, Australia

Chapman & Hall India, R. Seshadri, 32 Second Main Road, CIT East, Madras 600 035, India

First published in 1995 as a special issue of the Chapman & Hall journal
Biodiversity and Conservation, **4**(8)
First book edition 1996

For details of how to subscribe to *Biodiversity and Conservation* please contact the Subscriptions Department, ITPS, Cheriton House, North Way, Andover, Hampshire SP10 5BE, UK, tel: +44 (0)1264 342713, fax: +44 (0)1264 364418, or visit the journals internet site at http://www.thompson.com/journals.html

Visit the Biodiversity Resource Center at http://www.biodiv.com/biodiv.html

Printed in Great Britain at the Alden Press, Osney Mead, Oxford

ISBN 0 412 70980 5

A catalogue record for this book is available from the British Library

Library of Congress Catalog Card Number: 96-83710

Contents

Contributors

John P. Barkham
School of Environmental Sciences
University of East Anglia
Norwich
NR4 7TJ

Andrew Brennan
Department of Philosophy
The University of Western Australia
Perth 6907
Australia

Bob Carling
Senior Commissioning Editor
Life Sciences
Chapman & Hall
2–6 Boundary Row
London
SE1 8HN

Gary L. Comstock
Bioethics Program
Iowa State University
403 Ross Hall
Ames
Iowa 50011-2063
USA

Nigel S. Cooper
The Rectory
40 Church Road
Rivenhall
Witham
Essex
CM8 3PQ

Calvin B. DeWitt
Au Sable Institute Outreach Office
731 State Street
Madison
Wisconsin 53711
USA

Graham Dutfield
The Working Group on Traditional Resources Rights
Oxford Centre for the Environment, Ethics and Society
Mansfield College
University of Oxford
Oxford OX1 3TF

David R. Given
101 Jeffreys Road
Christchurch
New Zealand

Alan Holland
Department of Philosophy
Furness College
Lancaster University
Lancaster
LA1 4YG

Jane M. Howarth
Department of Philosophy
Furness College
Lancaster University
Lancaster
LA1 4YG

Eddie T. Idle
English Nature
Northminster House
Peterborough
PE1 1UA

Kristina Plenderleith
The Working Group on Traditional Resources Rights
Oxford Centre for the Environment, Ethics and Society
Mansfield College
University of Oxford
Oxford OX1 3TF

Darrell A. Posey
The Working Group on Traditional Resources Rights
Oxford Centre for the Environment, Ethics and Society
Mansfield College
University of Oxford
Oxford
OX1 3TF

Susan Power Bratton
2225 Parkside Drive
Denton
Texas 76201
USA

Rory J. Putman
Department of Biology
University of Southampton
Boldrewood Building
Bassett Crescent East
Southampton
SO9 3TU

Brent Waters
Omer E Robbins Chaplain to the University
University of Redlands
1200 East Colton Avenue
PO Box 3080
Redlands
California 92373-0999
USA

Foreword

Writings on environmental ethics are ten-a-penny nowadays. But analysis of the ethical implications of applied ecological science is rare. This publication seeks to fill that space. Ecologists understandably distance themselves from the more aggressive environmentalists, believing the job of the ecologist as scientist is to understand more completely how natural systems work. Most practising ecologists apply scientific ethics to their research methods and interpretations. That is, they follow established rules of scientific procedure, such as replication, blind testing, control variables and the examination of null hypotheses. In laboratory work, meticulous care is taken to ensure that basic techniques of measurement and comparison are also followed. But it is not so common for practising ecologists to be fully aware of the broad ethical envelope that surrounds their work. And when they are made aware, the shock of realizing how value-laden and socially-constructed is the scientific method can be quite intense.

For example, in the UK, the National Trust was pressured by its more vociferous members to ban the hunting of deer, and eventually foxes, hares and other quarry on its property. In the modern spirit of animal rights and increasing popular antagonism to hunting, the Trust could not afford to ignore these calls without losing membership income. So it asked a distinguished wildlife ecologist to conduct research into how far deer are stressed, and conceivably might suffer pain, when being pursued. Immediately there was an outcry that this was a stalling tactic. The Trust, badgered by the anti-hunting lobby, was genuinely unsure whether to stop the killing, so it set up a piece of research that it thought would prove the issue 'scientifically'. The political climate in which this work was to be undertaken was already antagonistic before the work has even begun. Whatever the results, both the method and the conclusions are bound to be fiercely criticized.

The above study, and a similar study of the implications of foxhunting on fox populations and local economies, raises the issue of how far the professional ecologist should specifically become involved in matters of ethical controversy. Equally important is the issue of how far ecological research can throw light on ethical dilemmas and help to re-establish them in the public mind.

These are the issues addressed by this collection of papers. One line of analysis is that the values embedded in any scientific method do not always implicitly embrace the significance of human relations to the natural world. Jane Howarth argues for the incorporation of phenomenological approaches to the study of ecology, namely the description of the world as experienced, with all its mysteries and emotional stimulations, as opposed to the world as revealed by scientific scrutiny – a world as she sees it of 'mechanisms and causal regularities'. This would suggest, as Cooper indicates, that the ecological management of churches and graveyards should take into account the continuity of human life, and the mood of light and shade that such environments carry. Simply caring for the fabric of the ecclesiastical building without noticing the mosses and liverworts that have become an integral part of that fabric is no longer sufficient. Ecologists can reveal these interdependencies, and if Howarth is correct, suggest the best combination of nature conservation and cost-effective building maintenance.

Putman addresses a similar theme when considering the relationship between bird ringers and animal taggers and the feelings of the animals involved. Any code of practice has to be believed in, with emotional conviction, it should not just be regarded as a rule of good field research. In general Putman believes that this is indeed the case for most of his colleagues. His argument, however, raises the all important issue of the ethical component in laboratory and field research techniques, and the extent to which this becomes part of the personal outlook of the researcher rather than any set of professional principles.

Brennan raises the contentious theme of the use of economic principles in the judgment of environmental values. His critique, namely that the very act of valuation through surrogate market procedures demeans the attachment of the human spirit to the natural world, is becoming recognized by thoughtful economists. There is a zone of convergence between economics, ecology and ethics that is being explored in the form of participatory ecological economics. Here the functions of ecosystems are being understood in terms of both their intrinsic worth as well as the cost of replacing them by artificial means. The two sets of information are then presented to various interested parties to thrash out common positions through such devices as mediated dispute resolution. This is likely to become an important new area for interdisciplinary enquiry.

This collection is almost unique in its orientation, and certainly thought-provoking in its argument. It calls for the ecologist to be both professional and personal, to adopt the rules of the discipline and to be prepared to change these rules as both popular and personal values change. It also calls for a renewed sense of awareness of the sensitivities of many to the rights of existence of many species, to the overriding importance of increasing biodiversity as a symbol of solidarity and commitment to healthy ecosystems, and to the willingness to be accountable to a wider range of social and political opinion when researching and presenting results. Possibly not all practising ecologists are quite so sensitive to these issues as might be desirable in this modern age of participatory science. Hopefully more will become so, having read the essays that make up this imaginative and timely publication.

Timothy O'Riordan
School of Environmental Sciences, University of East Anglia, Norwich NR4 7TJ, UK

Preface

Ecologists and ethical judgements

> Nowadays, scientists have always to consider themselves as agents, not merely observers, and ask about the moral significance of the actions that comprise even the very doing of science. Toulmin (1982).

What is the moral significance of the very doing of ecology? This was the question that led one of us (Nigel Cooper) to convene the Ecologists and Ethical Judgements symposium at the Sixth International Congress of Ecology (INTECOL). The question brings two fields of enquiry close together and at least two different sets of people. The aim was to stimulate interdisciplinary reflection, to help ecologists to reflect on the ethical dimensions of their science, and to feed scientific insight to philosophers and theologians.

Of course, all sorts of possible questions could have been posed. It would have been very worthwhile, for example, to have asked how ecologists become caught up in political processes, to have made a sociological study of the role of ethics in ecology or to have asked what psychological factors influence our relationship to nature. Actually the symposium concentrated on philosophical and theological questions. Not all the contributions are published here. Other publications which arise from papers presented include Farnsworth and Rosovsky (1993), Lee (1995) and Howarth (1995). Elizabeth Farnsworth (1995) has given an account of the meeting.

What then might be the embedded ethics in ecology? This may seem a puzzling question for those convinced of the value-neutral stance of science. On this view science describes what is, ethical judgements may then follow. Jane Howarth surveys some of the analyses that do claim to find ethical presuppositions at the foundations of ecology. She offers one approach in phenomenology that acknowledges the values at the core of our relationship with the world, i.e. the world we live in rather than the world we reflect upon.

Much work has been done by ethicists in attempting to establish the moral basis for environmental care. Should we care for the world merely because it is of use to us, or are there entities out there which make a moral claim on us because of their intrinsic worth? The science of ecology would seem to influence this debate in environmental ethics through its description of the natural world and the putative entities such as communities and ecosystems that may inhabit it. Alan Holland's reviews some ways in which ecological theory may influence ethics. But Holland concludes that basing ethics on the latest scientific view makes it too much of a hostage to changes in science and that recognising our intimate relationship with the natural world might prove a better grounding for ethics. Gary Comstock criticises those ethicists who have accepted an organismic model of ecosystems because, as many ecologists have moved away from this model, these ethicists have been left stranded without scientific support. He proposes instead to extend traditional individualistic ethics to animals and, through them, to make the rest of nature of use to sentient beings.

Ecologists may well have come to their science through a love for nature but now feel forced to defend nature on economic grounds. To remain credible in a market-dominated culture it seems essential to appeal to the cost–benefits offered by conservation. At a practical level, economists and ecologists who try to quantify externalities and to get them incorporated into decision making are contributing to environmental protection. However, Andrew Brennan suggests that environmental ethics challenges this reductive notion of value; in fact, it challenges the entire Western moral tradition. Economism, he claims, may give a bureaucratic justification for the choices we make but it does not give a rational procedure for determining that choice in the first place. David Given describes the multicultural background to New Zealand's Resource Management Act which also aims to justify decisions but leaves unresolved the grounds of those decisions in the notion of 'significance', which remains undefined in the legislation itself.

Mention of laws and economics reminds us of the larger forces at work in society, among which science seems to be an overshadowed plant seeking its place in the sunlit canopy. Two of these larger forces are social injustice and human population growth. John Barkham argues that the loss of biodiversity will continue so long as there is economic inequity between North and South. Discouragingly he sees no sign that Northern electorates are ready to move towards greater social justice despite this being in their long-term interest. If there are to be major changes, all cultural forces will need to act together. Susan Power Bratton gives the example of restricting human population growth. She emphasises the importance of local factors and of religion. Ecologists and others concerned for the environment will increasingly look to the worlds' religions for support in bringing about the radical changes needed.

All religions are involved in this process, but because of its intimate relationship with Western society, science and technology, Christianity has sometimes been singled out for special blame as the root of our crisis. Whether this is so or not, it may play a crucial role in persuading people in cultures where Christianity is a significant force to care for what they believe to be God's creation. Calvin DeWitt describes the environmental injunctions of the Great Code of the Bible and Brent Waters' paper is an example of one of the many recent attempts to integrate environmental concerns with Christian theology.

The practical ecologist may wish to ask questions such as: what are the ethical issues around the release of genetically modified organisms, or how are species selected for special conservation efforts, or how much intervention is justified to save them. Rory Putman reviews the ethical issues involved in experimental procedures, particularly with respect to animal welfare. Eddie Idle describes the thinking behind the management of nature reserves. Nigel Cooper, in a paper written after the INTECOL meeting, illustrates the application of ethics to nature conservation in churchyards. The ecologist also faces decisions involving professional ethics such as the impartiality of a consultant's advice, or how just are the employment opportunities for ecologists. A recent concern has been the protection of the intellectual property rights of indigenous peoples. Darrell Posey, Graham Dutfield and Kristina Plenderleith set out some of the principles ecologists should adopt when working with such people with local knowledge.

It was very evident at the symposium that the two groups of contributors – the scientists and those studying the humanities – brought with them two different cultures. As a banal yet significant example, it was remarked upon that the scientists showed the obligatory slides to illustrate their talks while the philosophers just spoke without visual aids. We think the papers collected here show similar but more profound divisions. We are, after all, only at the start of learning from each other but we believe the papers show the promise of this fruitful intercourse.

One significant feature is that the two groups hold asymmetric expectations. Ethicists have been asking ecologists, 'What are the structures of the natural world and how do they work?' Fortunately for them, this is the same question as ecologists have been asking themselves and so it is relatively straight forward for the ethicists to pick up their ideas and to run with them. Ecologists are, mostly, much less interested in the questions philosophers ask, such as 'Is the natural world worth conserving?' They are predisposed simply to accept that it *is* worth conserving. Their concerns are more immediate and pragmatic: 'How do we apply all this theory to this particular tricky case?', and 'Where can I get the funding to continue my research!' Ethicists, on the other hand, use phrases and concepts that may mean something entirely different to ecologists. For example there is a good deal of scientific literature on the concept of the stability or fragility of ecosystems, which has obvious practical management implications. It would be obvious at this point to appeal for public policy on environmental issues to be based on more than just nebulous concepts of whether a particular ecosystem is stable/robust, or fragile/unstable. But practical and urgent decisions need to be made on matters of conflicting priorities without recourse to a sterile debate about the meanings of words. Nonetheless, concepts of stability/fragility (and so on) need to be made more explicit, because of their practical implications for management, and the analysis of stability/fragility (at local ecosystem level as well as at biome and global level) is an example of a potentially fruitful area of collaboration and further work in the interface between ethicists and ecologists.

This collection of papers is therefore a contribution towards both groups asking and answering each others' questions. We would like to think that we have stimulated those at the congress and you, the readers of this publication, to take this further.

We wish to thank all the contributors to this publication and those speakers at INTECOL whose work does not appear here. We are also very grateful to the organizers of INTECOL and all those who have advised us as editors.

References

Farnsworth, E.J. (1995) Oikos and Ethos: setting our house in order. *Trends in Ecology and Evolution* **10**, 56–7.

Farnsworth, E.J. and Rosovsky, J. (1993) The ethics of ecological field experimentation. *Conserv. Biol.* **7**, 463–72.

Howarth, J. (1995) The crisis of ecology. *Environ. Values* **4**, 17–30.

Lee, K. (1995) Beauty for ever. *Environ. Values*, **4**, 213–26.

Toulmin, S. (1982) *The Return to Cosmology: Postmodern Science and the Theology of Nature.* Berkeley: University of California Press.

Nigel Cooper
Rector of Rivenhall and Ecologist on the Chelmsford Diocesan Advisory Committee, The Rectory, 40 Church Road, Rivenhall, Witham, Essex, CM8 3PQ, UK

Bob Carling
Senior Editor, Chapman & Hall, 2–6 Boundary Row, London, SE1 8HN, UK
bob.carling@chall.co.uk

1

Ecology: modern hero or post-modern villain? From scientific trees to phenomenological wood

JANE M. HOWARTH

Department of Philosophy, Furness College, Lancaster University, Lancaster LA1 4YG, UK

This paper sets out to launch a challenge to the usual 'modernist' view of the relationship between ecology and ethics. Two 'post-modern' interpretations of this relationship are considered. The first 'deep' interpretation holds that ecology reveals that nature has intrinsic value. The second interpretation derives from the work of Michel Foucault. The aim of his critique is to reveal how certain values are taken for granted by the acceptance of certain scientific models, and how the acceptance of those models as true makes it impossible to articulate alternative values. I end by suggesting, contra Foucault, that phenomenological enquiry could serve to articulate alternative and perhaps more eco-friendly values.

Keywords: ecology; Foucault; phenomenology; Heidegger; Merleau-Ponty

Introduction

Ecology is a rapidly developing science, with many of the problems – technical and philosophical – which characteristically attend such developments. At the same time a great burden of responsibility is placed upon the science, for we face enormous problems concerning our natural environment and as a culture we tend to look to science to solve our problems. There is an alternative, if rather quiet, voice in our culture which, so far from looking to ecology to get us out of our difficulties, sees ecology and modern science generally – or at least modern faith in science – as largely responsible for getting us into these difficulties. What does this voice have to say and should we listen to it?

It is the orthodox view in our culture that the difference between scientific and ethical enquiry is that scientific enquiry seeks to reveal 'facts' about its subject matter while ethical enquiry addresses questions of 'value'. In particular, scientific ecology is concerned with facts about the components and workings of the ecosphere while environmental ethics is concerned with whether the ecosphere, its components and workings, are good or bad. Meta-ethical enquiry will ask whether goodness or badness are 'discovered' in the ecosphere, or 'projected' onto it by us. The results of these ethical enquiries aim to guide practice by determining what we ought to do and how we ought to behave with respect to the ecosphere. One important area concerns our use of the technology which scientific enquiry makes possible. Ethics aims to guide, e.g., what uses we make of technology which affects the environment, and also what technology we choose to develop.

It is part of this view that scientific ecology and the development of technologies are themselves value-free. Environmental ethics, in contrast, must not be fact-free. What we ought to do depends essentially on what we can do, and that depends on what our capabilities are and on what the facts are, for example, about the results of our activities. In that sense scientific enquiry is held to be primary, ethical enquiry secondary.

Is this orthodoxy correct? It is one which has been dubbed 'modernist' and subjected to a range of 'post-modern critiques'. The terminology needs explanation. The modernist view, deriving from the 17th-century philosopher Descartes, is that the world comprises two kinds of thing: matter – which operates causally, deterministically, mechanistically – and mind – which operates rationally, and alone has the capacity to value and so is, in that sense, the source of all value. Rational, scientific, enquiry can discover facts about how the material world works. These facts neither entail nor presuppose any claims about value. The 'post-modern' perspective is one that reflects upon modernism with the aim of revealing it, not as the universal truth it purports to be, but as the orthodoxy of a particular culture. 'Post-modern critiques' of modernism involve analyses of modernism which seek to undermine its claims to universal truth.

One strand in post-modern thought is concerned to re-appraise the status of scientific knowledge. One element in this re-appraisal is to challenge the alleged value-free status of scientific knowledge. The claim to be value-free involves two distinct theses. First, it is claimed that the realm which the natural sciences investigate is a factual realm. If values attach to nature, it is not the task of science to explore them: science investigates the factual character of the world, aims to discover facts about it. The second claim or ideal of value freedom involves, not the subject matter to be investigated, but the scientific investigator. The scientist does not import any values into the enquiry. He or she is a purely rational, value-neutral, investigator. Any attachments to, or interpretations or evaluations of, the subject matter which the scientist as an individual might hold are set aside when the scientist is engaged in his or her professional investigations.

Post-modernism challenges the modernist separation of facts and values. We can distinguish two post-modern critiques of scientific ecology in the light of these two ideals of value-freedom. One denies the first aspect of value-freedom; the other denies the second. In brief, the first critique – sometimes called the 'deep' ecology position – claims that ecology is not a purely factual enquiry for it reveals values inherent in or intrinsic to the natural world. The second critique holds that ecology is not value-free because it is geared to prediction, manipulation and control of nature. It thereby presupposes value: namely the utility or instrumental value of nature as a human resource. I shall explore these two interpretations in turn, and consider the modernist response to them.

'Deep' ecology

The 'deep' interpretation of ecology claims that scientific ecology reveals the ecosphere, and its component ecosystems, as possessing intrinsic value (Lovelock, 1979; Rolston, 1992). Deep ecologists have heralded ecology as a truly post-modern science which rejects the modernist separation of fact and value and discovers value in its subject matter. Ecosystems, they claim, are shown to be not merely highly complex mechanisms. They display features such as self-directedness, purposiveness, harmony and balance. Ecology, it is claimed, recognizes the holistic character of the ecosphere, its way of working as one single organism and not as a collection of mechanisms. They have suggested that ecology reveals that ecosystems are more than the sum of their parts. They have likened ecosystems to human communities, suggesting that balance and harmony are achieved by the members or components acting 'for the sake of' the whole. The components of ecosystems act in harmony with each other, even for the sake of the greater good of the system. Any such system is to be respected for its own sake, for its integrity.

Systems of energy

To flesh out this interpretation, look at how it would apply to two examples of models used in the study of ecosystems: systems of energy and systems of information. Ecosystems can be presented as systems of energy flow. Many ecosystems are extraordinarily complex so that they cannot be understood in the same way that simple non-living systems are understood. One popular way of posing the problem confronting the contemporary scientific ecologist is that ecosystems appear not to obey the second law of thermodynamics. The second law describes how energy transfers are always 'inefficient', involve energy 'waste' or increase of entropy. If increased entropy were a feature of ecosystems, then, it is claimed, they would tend to disorder; but, so the claim goes, they appear not to do so. They 'recover' from disturbances, display 'equilibrium', tend towards order and complexity.

The 'deep' interpretation would seize on this talk of 'recovery' and 'tendency to order' and take it to entail value. On this basis, it would interpret the behaviour of systems as 'recovering' or 'tending towards order' as showing that ecosystems are self-directing, teleological, having purposes, aims, goals which are to be respected. It might even be suggested that the ecosphere has something comparable to human consciousness, that its unpredictability is due to its having something comparable to human freedom.

Systems of information

The second model represents ecosystems as systems of information. The usefulness of such a model is clearly related to developments in information technology which have facilitated the handling of the highly complex ecological data. 'Deep' interpretations might take the use of this model in ecology as grounds for claiming that ecosystems have intelligence, that components are able to act together and for the sake of the whole because there is communication through the system. Nature is not, as Descartes had it, a mere mechanism; the complexity of the ecosphere is as great or greater than that of the human brain. If nature is to be understood on the model of a machine at all, the machine has to be an intelligent one: a sophisticated computer, the sort of model which can also be used to explain human consciousness and the workings of human societies.

'Scientific' ecology

The 'modernist' or 'scientific' response to such 'deep' interpretations ranges from incredulity through irritation to horror. Whichever, the 'deep' interpretation of ecology is rejected *in toto*. Ecologists, the scientific modernist would claim, may use language of balance, harmony, order, even, on occasions, purposive talk of ecosystems or their components. But this is talk: models, metaphors. Balance or harmony are not to be thought of on the model of social values such as justice or peace; the balance is a balance of the energy budget. Computer models are used to order highly complex data. The information is information for us; but the ecosystem itself does not 'understand' or communicate this information in any sense which supports the 'deep' interpretation. The use of information models does not imply intelligence in ecosystems: the researchers understand the information, the components of the ecosystems do not.

The models, the modernist would claim, are not to be taken literally. The task of ecology is to discover the mechanism by which ecosystems work, the patterns governing their fluctuations. It seeks to discover mechanisms which determine the behaviour of components, or regular patterns in that behaviour; it does not assume some mysterious communication and altruism. Even in complexity or 'chaos' theory, where unpredictability is acknowledged, there are no conclusive grounds for believing that the ecosphere does not operate deterministically. Unpredictability, where it occurs, might rather be due to the impossibility of gathering sufficient, or sufficiently precise, data.

There is, according to the modernist, no ghost, no intrinsic purpose, in the machine; the machine is an extraordinarily complex one, but it is nonetheless one which operates deterministically. If we are to find value in it or grounds for treating it one way rather than another then, modernism insists, we must look elsewhere than in the findings of ecology alone. Is this modernist defence of science as value-free, concerned only with the facts about how things work, accurate? A second post-modern interpretation would say it was not.

The influence of Foucault

This second interpretation claims that ecology is not value-free it is not a value-neutral enquiry because, in its methods, classifications, models and theories, it presupposes a set of values. These allegedly 'presupposed' values stand in sharp contrast with the allegedly revealed 'intrinsic' values of nature claimed by the 'deep' position; the values which, according to this second interpretation, ecology presupposes are instrumental values: utility and power; the presumption is that nature is a resource, it has utility for us, and ecological enquiry is based on an attempt to achieve power and domination over nature. The following interpretation is based on the work of the 20th-century French intellectual and post-modern thinker, Michel Foucault. Foucault was influenced by the work of Nietzsche who recommended that we should enquire not into the truth about value, but the value of truth. Foucault's central question, with respect to any branch of knowledge which he considered, was: what gives these claims the authoritative status of truths? He looked for answers to his question in the broad cultural and historical context in which the truths in question have come to be accepted. This context includes the intellectual but – and often more significantly – the practical. What past and present practices, Foucault asks, give rise to the 'truths', and what practices do the 'truths' make possible? In less contentious terms, Foucault's overall question is: what gives certain theories or models their explanatory power?

In broad terms, Foucault's line of argument is that the status attained by a theory as knowledge, as truth, and its explanatory power, depends not on its giving a full, accurate, factual description of how its subject matter is, but in its usefulness, in its capacity to give power over its subject matter. To this end scientific investigation selects, classifies and organises its data in a way which will reveal 'surfaces of power'. These surfaces of power are points at which we can intervene to change, manage and manipulate what is being investigated. The authority and explanatory power of scientific theory, its status as the truth, derives from the fact that it gives power over its subject matter. This quest for power Foucault regards as a value implicit in scientific enquiry, but masked as fact as long as scientific theory is given the status of value-free truth. Modernism, in ignoring this basis of truth in power, in presenting scientific truth as a neutral, interest or value-free account of how the world fundamentally works, covers up this value, thereby legitimising actions in accordance with it, actions of manipulation and control. It also marginalises other, alternative, voices which conflict with this authoritative one. One voice which it marginalises is that of the 'deep' ecologist who believes that scientific ecology reveals value in nature. Since this position shares with modernism the view that scientific enquiry does not presuppose value, it too is a target for the Foucauldian attack.

It is important to realise that Foucault is not accusing individual practitioners. They are controlled, 'constructed', by the system of scientific practices of which they are part. In support of this, one might cite the way natural scientists are trained: the discipline they must acquire, the detachment they must attain, the commitment to value-free enquiry which they must learn to reconcile with the need to select and deal only with useful data. Individual freedom and responsibility, like scientific neutrality, are features of the modernist world view which post-modernism challenges. For Foucault, it is institutional practice, science not scientists, which has authority and power.

A Foucauldian critique of ecology

Foucault was primarily interested in the human and social sciences, though he did turn his attention, on one occasion, to physical geography (Foucault, 1980). What emerges if we subject, as Foucault himself did not, the science of ecology to this kind of interpretation?

Look first at the cultural and historical context in which ecology has arisen. We live in a very science-oriented society: not, alas, in the sense that we are knowledgeable about science; but in the sense that we look to science, not only to tell us how the world works, but to solve our problems. Currently, we are facing, if not an environmental crisis, certainly a number of acute ecological problems such as pollution, species extinction, and climate change. We are coming to fear the results of our using the world's resources in a thoughtless way, to be concerned that we may render the planet uninhabitable at least by humans. Ecology ranks along with other sciences already accepted as authoritative, and it arises in a situation where the general feeling is that we face a potential environmental crisis about which we must do something. The Foucauldian claim would be that this is the social context in which ecology must be seen: great responsibility and great urgency attend ecology.

In the light of this pressure, ecology needs to predict potential disasters, so that we might seek to avert them; and it also needs to predict the consequences of our interventions in nature, so as to make possible sensible management. The status of ecology depends on its predictive powers. Because ecosystems are so complex, prediction involves selection in two respects. First, the ecologist must select from all the available data just those which are relevant to prediction. Second, the ecologist must select which predictions are the important ones to seek. The Foucauldian claim would be that both of these selections are performed with a view to how we can intervene: they are geared to reveal 'surfaces of power'.

This critique might cite, in its support, that ecology does not aim to predict with 100% accuracy, but only probabilistically. For example, in population dynamics, one needs only to predict the size of a population within a certain range, precise numbers are not required. Nor, in oceanography, would one aim to predict, e.g., the precise pattern of individual waves. This selectivity, the critique would claim, indicates that intervention not prediction is the ultimate goal. We seek to predict only where we see the need or have the capacity to intervene. Only if a population is dangerously high or dangerously low do we need to know and this does not require precise numbers. Only where wave power is threatening or useful do we need to know, and this will not typically require knowledge of individual waves. Predictions are geared to revealing 'surfaces of power', places where we can exercise power over nature.

Models are constructed with a view to revealing these surfaces of power. Information is selected or deemed irrelevant, according to whether it fits the model, and classified, interpreted or transformed so that it will fit the model. Ecology interprets, presents and reveals, nature as a set of surfaces on which we can exercise power, as an instrument which we can use, a resource for us. This, the Faucauldian critique would claim, is scarcely the neutral, value-free enquiry which modernism claims it to be. It is value-laden through and through. The presupposed value is instrumental value or utility.

The models, it would further be claimed, gain authority because we have the technology to enable intervention on the surfaces of power which these models reveal. Consider the two examples. The first is that of a system of energy flow. What, this interpretation asks, is the technological backdrop against which this model gains its plausibility? It would point to the massive energy industry, the great range of technology we have for harnessing and storing, not to say using, energy. Where the 'deep' interpretation focuses on the talk of balance, harmony, order, this interpretation would rather pick on how, in the energy industry, talk of energy naturally gives rise to talk of efficiency and inefficiency, of waste, of balancing energy budgets. The use of this model in ecology naturally brings with it these considerations of efficiency already embedded in energy technology. The energy-flow model highlights points of energy transmission, reveals, as surfaces of power, places where we can increase efficiency, tap into the energy flow, utilise what would otherwise go to waste. The model is geared to isolating those features which enable us to utilise nature's resources. There is no place in the model for features of nature not relevant to this end: such features, it is claimed, are 'rendered invisible'. The model excludes any representation of nature other than as a source of energy for us. The interpretation would further look at the technology which is facilitated by this model. It could cite the usefulness of such models in, e.g., intensive farming, in gaining maximum output in terms of human food from minimum input in terms of fertilisers.

The second model is that of a system of information. What is the technology which makes this model plausible? Clearly it is information technology, highly sophisticated computers. Because ecosystems are so complex, the data is manageable only by computers. Since what computers handle is information, it is natural to introduce information models of ecosystems. Where the 'deep' interpretation saw this as grounds for ascribing to ecosystems and their components, intelligence and the capacity to communicate, this interpretation would again present this in a very different light. The information model requires that we 'interrogate' nature, discover or force it to 'confess' to us, its (her?) 'secrets'. If this seems a trifle old fashioned, look instead at the way information technology regards information. Computers essentially serve our interests, they perform tedious tasks which previously only humans could do. One such task is accessing information. Computers succeed or fail according to how efficiently they do that, how user-friendly they are. The elegant program, no matter how it works, will be one which produces the required information in an intelligible form at high speed, and enables input with least effort on the part of the user. Information technology is geared to ease of access of useful information, and ease of manipulation and transformation of information. So with the information model in ecology, we formulate the questions, collect, classify and marshal data, not in the spirit of neutral enquiry, but with the aim of turning the information to our own use. This model, by presenting the relevant information in an accessible form, makes nature accessible to us, it reveals surfaces of power, points of transmission of information, where we can intervene to transform the system. Once again, nature is represented as an instrument. Like the energy model, this model gains authority because it facilitates the achievement of control and power over nature, and that goal of power is, on this interpretation, a value. That value determines all our subsequent understanding of nature and it is disguised, not recognized as a value, because the model is given the status, the authority, of the truth.

This interpretation would then ask: what technology does the information model make possible? What 'surface of power' does it reveal? It highlights the gene as an important information carrier. Intervention at that point can affect an entire ecosystem, just as misinforming a spy can redirect a government. Biotechnology enables us to manipulate, even create, components of an ecosystem so as to prevent blight to our crops or disease in our children. This sort of power is, the Foucauldian claim, would be what gives this model authority. The accuracy or completeness of the model are not what determines its status as truth, what matters is how many surfaces of power it reveals, how much power we can exercise at those surfaces, how much power it gives us over nature.

So, overall, it is the utility value of nature which dictates the character of the models or theories. The power of the models consists in their ability to render nature of use to us, to give us power, domination, control over nature; and it is this which gives them their authority, their status as scientific truths. This then masks the underlying value. The value, the utility of nature, is smuggled in under the guise of value-neutral fact.

Scientists might respond to this interpretation by agreeing that science aims to predict, even that this influences the character of their theories, that models are justified by their predictive powers. They might even concede that power over nature is in some sense a goal of their studies, and so a value they presuppose. However, they might go on to claim that this does not preclude, at the level of application of the theory, at the level of environmental management, alternative values which do not treat nature as merely a resource to be manipulated and controlled for human ends. They may reveal surfaces of power, but how power is to be exercised and to what end, is a different issue. We need not exercise power entirely for our selfish use.

But this would be to miss the point. Foucault's claim is not that knowledge gives power, but that whatever gives power attains the status of knowledge. Any interpretation or description which is not empowering, which does not reveal surfaces of power where we can intervene, will not gain that status. The effect of models or theories gaining authority is that they are taken to be the truth. One thing which this involves is the marginalization of alternative claims to truth. The authority given to the scientific models, presupposing as they do the value of utility, essentially problematises other values, marginalises discourses in which other features of the world can be talked about, other values can be expressed and discussed. One might cite in support of this how, in modern society, the dominant attitude towards values which are not utility values, is that they are mere personal preferences, subjective opinions with no authority beyond that of the individual who holds them. Equally, our culture dismisses non-scientific talk about nature as 'mystical' 'romantic' or 'sentimental', as again mere subjective interpretation. We may talk about values not based on power, talk about world views which do not reveal surfaces of power; but they will never gain authority, they will always be 'alternative' just because they do not offer the prospect of power.

The modernist response to this critique would be much the same as its response to the 'deep' interpretation: i.e. this critique has picked on models, metaphors, talk, which may appear to embody certain values, but which are peripheral to the overall enquiry. Scientific enquiry might reveal surfaces of power, but that is not its primary goal. Technology may enable intervention at these surfaces; but the technological intervention does not determine the structure of the theory. The theory aims to reveal how the world actually works, not just how it can be made to work for us. They might add that, where ecology does facilitate intervention, e.g., to save a species from extinction or to eliminate a pollutant, it would surely be irresponsible not to do so, or not to continue with the study which facilitates it.

A different response is possible. One might present a challenge to the Foucauldian position. If, as it claims, scientific accounts of nature exclude other accounts, and if, by implicitly endorsing one set of values, they exclude other values from serious consideration, then is not the burden on Foucault to seek to articulate, to seek to gain, if not authority, at least currency, a voice, for these marginalized alternatives?

The phenomenological approach

Foucault himself would probably not have risen to this challenge, at least if it is interpreted as a challenge to describe the world and the value attaching to it 'as it really is'. For that, we must turn to a different school of philosophy, phenomenology, by which Foucault was influenced but whose positive programme he rejected. The phenomenological critique of science would be along somewhat similar lines to the Foucauldian one: that science involves the adoption of a certain stance towards its subject matter, the stance may be justifiable for certain purposes, to achieve certain ends; but it is an error to suppose that its 'truths' reveal how the world fundamentally is.

The phenomenological task is twofold: to 'strip away' theoretical and everyday assumptions and presuppositions about the world and our place in it, and to offer presuppositionless descriptions of the 'phenomena': how the world reveals itself to us, if we allow ourselves to experience it free from any distorting presuppositions about how it must be. Phenomenology claims to reveal, thereby, significance, meaning or value in the world; features which scientific explanations ignore.

Phenomenologists point out how different the world as we experience it – what they call the 'life-world' (Husserl, 1970), the 'lived-through-world' (Merleau-Ponty, 1962) or 'ready-to-hand' world (Heidegger, 1962) – is from the world as science depicts it. The 20th-century German phenomenologist, Martin Heidegger (1962) writes:

> But when this [the scientific investigation of nature] happens, the Nature which 'stirs and strives', which assails us and enthralls us as landscape, remains hidden. The botanist's plants are not the flowers of the hedgerow; the 'source' which the geographer establishes for a river is not the 'springhead in the dale'. (p. 100)

Heidegger's French contemporary, Merleau-Ponty (1962), writes in a similar vein:

> To return to the things themselves is to return to that world which precedes knowledge, of which knowledge always *speaks*, and in relation to which every scientific schematization is an abstract and derivative sign-language, as is geography in relation to the country-side in which we have learnt beforehand what a forest, a prairie or a river is. (p. ix)

Further, on the difference between a scientific stance and everyday experience, Merleau-Ponty (1962) writes:

> Natural perception is not a science, it does not posit the things with which science deals, it does not hold them at arm's length in order to observe them, but lives with them; ... (p. 321)

> The sun 'rises' for the scientist in the same way as it does for the uneducated person, and our scientific representations of the solar system remain matters of hearsay, like lunar landscapes, and we never believe in them in the sense in which we believe in the sunrise. (p. 344)

In this context, the phenomenological claim would be that our ordinary experience of nature is not as ecology presents it. We do not experience, for example, it as an energy flow or a system of information.

Phenomenology seeks to draw out and articulate the meaning, the significance of the 'lived-world'. It is not, as modernism claims, made up exclusively of matter operating mechanistically. The lived-world has 'intelligible form': it makes sense to us, it is not a mere jumble. But the *way* it makes sense does not necessarily accord with scientific classifications. One sees a pond as teaming with life, a sunset as vibrant, a wood as dank and mysterious. We experience in nature all manner of properties other than those of interest to scientific classifications. Also, we detect in the world relationships between things which are not of relevance to a causal, deterministic explanation. The relations we detect in the world are much more intimate than causal relations. They would include such features as the harmonies of waves gently lapping a shingle seashore, the murmuring of the wind in the trees, the battle of a blasted tree to survive on the windswept hill. We learn to read the weather and the seasons. The clouds 'mean' rain, the falling leaves 'signify' the winter to come; birdsong 'heralds' the dawn. Deterministic accounts of these relations do not capture the harmonies, the resonances, the intimacy of relations between things in the natural world as we live in it. In our lived-worlds, we are perhaps more artists than scientists, or maybe it is rather that the artist seeks to capture features of the lived-world. Merleau-Ponty (1962) writes: 'I experience the unity of the world as I recognise a style' (p. 327).

Phenomenologists also emphasise the interrelatedness of subjects and the world. The 'objective', detached, rational stance of the scientist is not our fundamental way of being in the world. We are not, as modernism claims, logically distinct from the material world; nor are we, as subjects in that world, primarily pure rational consciousnesses. We are 'internally' related to the world: we are as we are because of how we relate to the world. To be sure, we do depend on nature, contingently, as a life-support system, a supply of energy which we happen to need in order to survive. But, what phenomenology points out is how much our lives, our patterns of behaviour, our skills and habits, are moulded by and to our environment. The context 'calls forth' the habit, the skill is inseparable from its context. We develop, as Merleau-Ponty puts it, 'sedimented' into our bodies, a wealth of knowledge and competence. The knowledge is not theoretical, but practical: we know how to – we can – do things in the world. This know-how can involve enormous implicit sensitivity to the world. The theorist, with all the computer power at his or her disposal, does not know, cannot predict, the individual flow of a wave; the successful surfer, in contrast, knows how to ride it. Presupposed by any intellectual comprehension is a practical, bodily understanding, an adaptibility and capacity to live in a world which has significance for us. Merleau-Ponty (1962) emphasises the relation of meaning between us and the world:

> ... the whole of nature is the setting of our own life, or our interlocutor in a sort of dialogue (p. 320).

The task of phenomenology is to describe this lived-world, with all its mysteries and significances, in contrast with the world as it reveals itself to scientific scrutiny, a world of mechanisms and causal regularities.

Phenomenology and ecology

This brief glimpse at the aims of phenomenology should be enough to see that a phenomenological account of the natural world will be very different from an ecological one. Phenomenology purports to describe the world as it is encountered pre-theoretically. It does not use models, it does not look for mechanisms, or seek to represent patterns as algebraic equations. It seeks rather to describe what the theories and models are theories and models of.

What, though, it might be asked, is the relevance of phenomenology to ecological science? There are three possible responses to this. First, if – as Foucault believed – it is true that science is concerned essentially with domination and control, then it is possible that some of the features of the lived-through world which phenomenology reveals are of no relevance whatsoever to ecology, for phenomenology does not aim to describe the world with a view to prediction, control or management. Indeed, it could be argued that to expect phenomenological descriptions to be of use is precisely to miss the point. A central aim of phenomenology is precisely to draw our attention to features of the world which do not relate to our interests in predicting and controlling it. Even where phenomenology seeks to describe our use of things for practical purposes, this is sharply distinguished from the theoretical descriptions of the scientist, whether or not these are construed as geared to prediction and control.

Alternatively, phenomenological description might focus on phenomena which are relevant to scientific theory, as confirming or challenging a theory. It is a familiar point that theory can influence perception; the phenomenologist might detect something relevant to a theory which a theory-orientated scientist might have missed. Phenomenological description might have some constructive role to play in the scientific enterprise.

But where the impact of phenomenology would perhaps be greatest would be in the application of the findings of ecology to the management of nature. For, if phenomenology is correct in its claims to find significance in nature and our relations with it, and to locate the fundamental significance of our lives in those relations, and if it is also right in its claim that science does not deal with those significances, then clearly the scientific predictions about the natural world and the results of our intervention in it, will also ignore those significances, and so would be incomplete. It would be for phenomenology to point them out.

It may take a phenomenological description to reveal how, for example, a radical change in the style, atmosphere, character of a piece of woodland might result from a change in biodiversity or the numbers of a particular species which are not regarded as ecologically significant because nothing is threatened. Phenomenology might reveal how a windfarm can change the entire character or significance of a landscape. Prior to the advent of windfarms, Heidegger (1993) wrote:

> The revealing that rules in modern technology is a challenging, which puts to nature the unreasonable demand that it supply energy which can be extracted and stored as such. But does this not hold true for the old windmill as well? No. Its sails do indeed turn in the wind; they are left entirely to the wind's blowing. But the windmill does not unlock energy from the air currents in order to store it. (p. 320)

Phenomenological description could also bring to awareness, not only our own lived-world, but the lived-worlds of other creatures. It might offer, to those who object to certain intensive farming methods, a way of articulating their worry. Free-ranging farmyard hens have a lived-world; battery hens do not, they only have life-support systems. The battery hens have been reduced to what the energy-flow model in ecology presents them as being; points of energy transmission. There may also be a comparable objection to be made against biotechnology. Merleau-Ponty (1962) wrote:

> Animal behaviour aims at an animal setting ... If we try to subject it to natural stimuli devoid of concrete significance, we produce neuroses (p. 327).

Overall, what phenomenology aims to do is to describe our pre-theoretical engagement with the world. It would claim further that it is in this engagement that values have their source. In interacting with the natural world, we come to value it: the intimacy of the interrelatedness makes it impossible for us not to be attached to it, to value it, for it is our home. In learning to read its significances, its harmonies and discords, and also its sometimes total otherness from us, we are engaging in activities which are the foundation of valuing. Valuing is an activity; we forge, develop our values, our attachments, by our interactions. This is what is fundamental to values and valuing. Values are neither discovered nor projected, they are, first and foremost, lived.

Does phenomenology bear upon how ecology and ethics are, or ought to be, related? If phenomenology is right in claiming that values have their source in our fundamental interactions with the world and that ecology abstracts from this, then that must surely cast doubt upon the orthodox picture of ecological enquiry as primary and ethical enquiry as secondary. For once we have, in the name of scientific enquiry, denuded our account of the world of those features which are the source of our valuing it, it is hard to see how we can either discover in or project onto such an account of the world any realistic, well-grounded values. Any attempt to re-instate value based on a model of the world which sets out to exclude it will fail to do justice to those values, for it has obscured their basis. Having extracted the juice from the lemon, liquid poured over the skin gets no purchase, it does not restore the original fruit.

Ecology abstracts from the lived-world. In so doing, phenomenology would claim, it ignores the very point at which value takes hold, i.e. the lived-world. That is the sense in which ecology is value-free: it begins by setting values aside. It describes the world denuded of those features with which we come to form attachments, which we grow to know, love and respect. The way in which ecology describes, classifies, presents the natural world is different from the way we need to characterise it as the valued world, the lived-world. We value nature as our home, our dwelling, we get attached to, come to value, particular familiar parts of it. We value the pond or the hedgerow as pond or hedgerow not as ecosystem, energy flow, or information network.

For phenomenology, the modernist picture of a sharp division between objective fact and subjective value is an abstraction from the lived-world where subject and objects, facts and values, are inseparably linked. The ethical theorists who ask whether values are in the object or in the subject miss the point; values lie in the essential interrelatedness of subjects and objects. In dividing up the world into subjects and objects, modernism makes values problematic. In focusing exclusively on objects, modern science further obscures from view the source of value.

This may not be a criticism of science. It may be more a criticism of those other institutions in our culture which listen only to the voice of science and relegate questions about significance and value, and their answers, to the realm of the merely subjective preference and opinion. Merleau-Ponty (1962) invokes us to listen to another voice:

> The whole universe of science is built upon the world as directly experienced, and if we want to subject science itself to rigorous scrutiny and arrive at a precise assessment of its meaning and scope, we must begin by reawakening the basic experience of the world of which science is the second-order expression. Science has not and never will have, by its nature, the same significance *qua* form of being as the world which we perceive, for the simple reason that it is a rationale or explanation of that world. (p. viii)

But if we are to hear the alternative, phenomenological voice, we must begin by adopting towards the world, an attitude which may well be the driving force behind scientific as well as phenomenological enquiry, what Merleau-Ponty, quoting Fink, calls ' "wonder" in the face of the world' for, Merleau-Ponty continues, 'it alone is consciousness of the world because it reveals that world as strange and paradoxical.' (p. xiii).

Acknowledgements

I would like to express my gratitude to the editors, Nigel Cooper and Bob Carling, for their valuable comments on an earlier draft of this paper. I am most especially indebted to my colleague, Michael Hammond, with whom I discussed the earlier draft in the light of those comments. That discussion issued, eventually, in a transformed and, I hope, much better paper.

References

Foucault, M. (1980) *Power/Knowledge* (C. Gordon, ed.) Sussex: Harvester Press.

Heidegger, M. (1962) *Being and Time, 1st edn.* (J. Macquarrie and E. Robinson trans.) Oxford: Basil Blackwell.

Heidegger, M. (1993) The question concerning technology. In *Basic Writings, 2nd edn.* (D.F. Krell, ed.) pp. 30–41. London: Routledge.

Husserl, E. (1970) *The Crisis of European Sciences and Transcendental Phenomenology* (D. Carr, trans.) Evanston, IL: Northwestern University Press.

Lovelock, J. (1979) *Gaia: A New Look at Life on Earth.* Oxford: Oxford University Press.

Merleau-Ponty, M. (1962) *Phenomenology of perception, 1st edn.* (C. Smith, trans.) London: Routledge.

Rolston III, H. (1992) Challenges in environmental ethics. In *The Environment in Question* (D. Cooper and J. Palmer, eds.) pp. 135–46. London: Routledge.

2

Ethics, ecology and economics

ANDREW BRENNAN
Department of Philosophy, The University of Western Australia, Perth 6907, Australia

This paper describes the general structure of an environmental philosophy. There can be many such philosophies, and those with their roots in economic theory have been extensively studied recently. Specific examples cited in the paper include the work of David Pearce and Robert Goodin. Economics-based philosophies can founder on the issue of externalities and a misplaced attempt to provide a comprehensive approach to valuing nature as a bundle of goods and services. It is argued that it is dangerously easy to slide from considering nature as a standing reserve of processes and objects that have the potential to satisfy human desires to the idea that it is nothing more than that. In general, the consequentialist basis of economics limits its usefulness in contributing to informed environmental decisions. But there is room for a sensitive use of institutional environmental economics as a partial guide for our reflections and choices. Any such development should take on board the existence of a plurality of perspectives on fundamental issues and the pluralism of values that can be found within moral theory itself.

Keywords: environment; ethics; economics; pluralism; Goodin; Pearce; values; cost-benefit analysis

Environmental ethics

The development of environmental ethics has been one of the most exciting and challenging things to happen in recent philosophy. To explain why this is so, it is necessary to understand the way that appeals to *community* and *commonality* have figured in western ethical theory, particularly since the Enlightenment. Although the eighteenth century Scottish philosopher David Hume remarked that it was 'evident' that 'beasts are endow'd with thought and reason as well as man' (1739), this did not lead him to view non-human animals as worthy of moral consideration. Indeed, it was not until 50 years later that Jeremy Bentham suggested that any being capable of suffering has a moral claim on us. For him, the morally important question to ask was not 'Can they *reason*? nor Can they *talk*? but, Can they *suffer*?' (1789). If once we concede that non-human animals can suffer, then it seemed clear to Bentham and his followers that we are obliged to take that suffering into consideration when making moral decisions.

The development of active movements against animal experimentation and intensive farming has been helped by considerations like Bentham's. If we humans share some common features with other animals – like the capacity for suffering – then this seems to give a basis for taking account of animal suffering when deciding what to do. People associated with animal rights and animal liberation movements have been quick to point to commonalities between human and non-human animal life as part of their case for political and social change. The capacity to feel pain is not the only commonality. Those fighting to preserve the cetaceans often build their case on considerations relating to the high intelligence of dolphins, or the social instincts, monogamy and nurturing behaviour of the great whales (Dobra, 1986).

The search for features common to humans and other animals is related in an obvious way to the attempt to build a *moral community*. For many ancient moral traditions, whether the Stoics of ancient Greece or the Confucians in China, the growth of moral maturity is associated with recognizing the moral claims on us of others outside what the Stoics called 'the community of our birth'. Most people have no difficulty in recognizing the moral bond between parents and children, the obligations associated with broader family life and the appropriateness of doing what is right by our friends, relatives, neighbours and so on. The terminus for this extension has traditionally been the human community as a whole. As we extend our moral cares beyond friends, family and people in our immediate neighbourhood, we naturally look for features which give a foundation for this extension.

In much of modern western philosophy since Descartes the grounding of moral respect for other human beings has been based on the following idea: each one of us is a centre of reflective conscious life; each has projects, desires and things we value; now, if I claim moral respect for myself on the grounds that I am a centre of values, desires and self-conscious reflection, then I cannot in all consistency deny giving moral respect to any other creature which is like me in these ways. From this idea flows a number of others. One is to do with equality. Being a subject of a centred, conscious life, with interests of the sort that humans have is not something which comes in degrees. No doubt some people are more intelligent than others; and some have richer sets of desires and projects than others. But all normal adult humans have some desires, some projects, some capacity for self-conscious reflections. So all normal adult humans are equally moral agents, and they are equally moral patients (that is, their interests ought to be taken into consideration by other moral agents).

Suppose we accept the line of argument just given. A moral community, linked by reciprocal recognition and obligation, is one founded on common features like the capacity for self-awareness and reflection. Now think about Bentham's argument for extending moral concern to other animals, an argument based on appeal to the capacity for suffering. What Bentham can be thought of as saying is that the capacity for suffering is just as important as the capacity for self-awareness and reflection when it comes to establishing a moral community. For the presence of that capacity in other animals gives a foundation for including them in a single moral community along with normal adult human beings. Modern theorists have even focused on the differences between children, people in comas or the confused elderly on the one hand, and healthy and intelligent primates on the other, to argue that our moral responsibilities towards some groups of human beings are less clear than our responsibilities towards at least some other animals.

So the ideas of commonality and community go hand in hand. Leaving aside the numerous possible objections to the style of argument just described, we could say that, for the most part, the extension of moral consideration to non-humans has been argued for on the same kind of basis that traditionally was used to support the extension of consideration from one group of humans to other groups of humans. In arguing that we should care, morally speaking, about the fate of other animals, Bentham and his followers are arguing in the same way that Aristotle could have argued (but, notoriously did not!) for including women as well as free men in the moral and political community.

It is against this background that environmental ethics is so challenging. Whereas arguments on behalf of non-human animals urge the extension of moral consideration in a perfectly traditional way, environmental ethics challenges the entire tradition. For environmental ethics takes seriously the idea that trees, flowers, rocks, rivers and even ecosystems deserve moral consideration from us. But when we look at what might ground the extension, then – although there are still plenty of common properties to choose from – it is hard to see that any of these is morally relevant. Some theorists have suggested that life itself is a foundation for moral respect (Naess, 1973, 1989; Taylor, 1986). But rivers and ecosystems, though supporting communities of life, are not themselves alive. Ecocentric ethics has argued that ecosystems require respect because they provide a location for life as well as producing it and permitting the development of new life forms (Rolston, 1988). But an argument of this sort moves away from any explicit dependence on features *common* to human beings, rivers and ecosystems. Abandoning the relationship between commonalities and community has posed a significant challenge to the mainline traditions in modern moral theory. In this way, environmental ethics has joined with feminism and the renewal of interest in Greek moral theory to challenge the dominant ethical paradigms of the last 300 years.

At times of challenge and intellectual crisis people move in many directions. Here are three trends that can easily be identified from a study of the literature. First, some thinkers embrace the new thinking even at the cost of inconsistency. Others illustrate a second trend by repeatedly changing their minds. A third trend, and one on which I will focus in the present paper, is what might be called the *reductive* one. It rejects the challenge posed by the new ideas and clings determinedly to old ideas, sometimes claiming that the existing paradigms already have the necessary resources for capturing what has been (wrongly) claimed to be a new, radical view. I will have more to say about this third trend later in the paper.

At such times of change, perhaps one of the most useful things a philosopher can do is consider the different positions or philosophies at stake so that at least some of the disputed territory can be mapped out. In the next section, I will briefly sketch a framework within which we can think about the competing positions. Then I will briefly discuss some recent work which has rejected the radical challenge of environmental ethics as I have outlined it so far. Finally, I will make some tentative suggestions about how those who are hostile to the non-anthropocentric drift of environmental ethics can start to take on board the challenge posed by non-anthropocentric theories.

Environmental philosophies

In one fairly popular sense of the term, a 'philosophy' is a general theory which explains or justifies actions, policies or positions. A company can have a sales philosophy, a political party an electoral philosophy and individuals can discuss their philosophies of life without reference to any academic work in philosophy. In this popular sense, there are many environmental philosophies. For present purposes, an environmental philosophy can be defined as a general theory linking humans, nature and values. Here is what a systematic environmental philosophy should contain:

(i) A theory about what nature is, and what kinds of objects and processes it contains.
(ii) A theory about human beings providing some overall perspective on human life, the context in which it is lived, and the problems it faces.
(iii) A theory of value and an account of the evaluation of human action with reference to the two points above.

(iv) A theory of method, indicating by what standards the claims made within the overall theory are to be tested, confirmed or rejected.

Much of the last 20 years' work has not been concerned with producing environmental philosophies. Rather, writers and theorists have developed fragmentary approaches to this larger task. For example, a great deal of attention has been focused on the question of values, sometimes without much explicit treatment of the other issues. This is not surprising, given our tendency to hold parts of the background constant while we tackle a particular problem. However, the challenge remains of articulating environmental philosophies which coherently link all of the above components. Some help with meeting this challenge comes from the work of those philosophers, such as Spinoza or Heidegger, who have already made some attempt to link humans, nature and values, and some recent work in environmental ethics has made reference to them (Zimmerman, 1983, 1993; Mathews, 1991).

What I have previously called 'frameworks of ideas' also contain, or at least permit the formulation of, such philosophies (Brennan, 1988). Christianity, for example, can be thought of as a framework of ideas relating humans, nature and values in a significant and interesting way. As an evolving tradition rather than a fixed set of doctrines it permits the formulation of several different environmental philosophies. For example, most varieties of Christianity are clear on the situation of humans, the problems they face, and the ways of overcoming these problems through adherence to faith and by appropriate actions. However, the account of nature and its processes can be given different, but equally Christian, formulations. For example, two Christian theorists could disagree over whether humans should be respectful stewards of God's creation or instead use nature as a mere resource, to do with as we please.

Not surprisingly, there has been a tendency to try to reduce the impact of the challenge made by environmental ethics as I have so far described it. Why should we bother to devise new environmental philosophies, after all, if traditional human-centred thinking can capture all the concerns of conservationists? Maybe the concerns that wild nature be preserved, that species be allowed to flourish, that non-human creatures should be respected for their own sake, not for ours – perhaps these can all be reduced to concerns about human welfare, the enjoyments and preferences of present and future human generations and the context required for sustainable and worthwhile human societies. Precisely this move has been made by some theorists influenced strongly by the 19th-century traditions of liberalism and utilitarianism or impressed by the power of economic theory and its depiction of human nature.

Green politics and environmental philosophies

Green political groups are often separated into the 'realos' and the 'fundies', or the light green versus the dark green (Dobson, 1990). Looked at in one way, these divisions reflect no more than the distinction between Gifford Pinchot, advocate of utilitarian wise use, and John Muir, whose reverence for the works of God and nature requires us not to intervene unnecessarily in natural systems. Looked at another way, as Bryan Norton has argued, perhaps the light greens focus on the realities of political action while the deeper greens are concerned more with the character of the agent (Norton, 1991). Here, then, is a parallel with a well-known idea in moral philosophy: the difference between agent-centred accounts of the good and a different conception focused more on states of affairs and their value.

Any decent moral theory has to find a way of connecting these two kinds of considerations. For when we reflect on our moral situation we are concerned to do the right thing and we are also concerned that we ourselves are – or try to become– agents for whom doing the right thing comes naturally. Of course, I do not keep my promise to meet my friend for lunch solely out of a concern to be the kind of person who keeps promises! To reflect too much on one's own character development during the course of moral decision-taking would run the risk of being too self-concerned and too little concerned about others (this is sometimes given as a problem for positions like that taken in Murdoch, 1970 and Nussbaum, 1986). Yet a concern with what kind of character I have and what kind of consistency – or otherwise – is present in my life and decisions is a perfectly appropriate moral concern. In an environmental matter, I may urge that we protect a forest because of the value of the trees and other animals in themselves; yet I may also reflect on whether a good life for human beings is one in which we recognize the intrinsic value of trees. In other words, a developed environmental ethic will have room for an account of two things: on the one hand, agents – and their characters, style of action and so on – and, on the other, the values which direct and guide action.

This last distinction, between green agency on the one hand, and green values, on the other, has been discussed recently by Robert Goodin (1992). His analysis of the problems faced by green politics draws upon a pair of distinctions. First, he separates green values from what he calls the green theory of political agency. The values of preservationism and respect for nature, he argues, are not essentially connected with a commitment to participatory democracy and decentralized organization. Green political theorists, he maintains, are mistaken when they insist that a central part of the green platform has to be associated with a renewed commitment to democratic and participatory processes, to devolution of decisions to groups of a humanly meaningful size and to local action. Likewise, Goodin uncouples another pair of ideas often held by green theorists to go together. To advocate green policies, he suggests, need not involve us in making immediate changes to our own personal lifestyles. Like Mark Sagoff, who admits to driving a car which leaks oil while displaying an 'environment now' sticker (Sagoff, 1988), Goodin suggests that we distinguish the domain of the personal from the domain of the public. (Note that Sagoff's associated idea of distinguishing the *agent as consumer* from the *agent as citizen* has been taken up by some economists, see Common *et al.*, 1993.)

Green political theory as standardly proposed thus commits a double error, according to Goodin. His analysis leaves it open for green politicians to argue for strong centralist remedies, the overriding of individuals and small communities, and the recourse to acts of aggression and violence in support of the protection of the environment. Indeed, if we try to articulate a position which might provisionally be labelled 'eco-fascism' (because of its structural similarity to European fascism of the 1930s), then it would appear that, for Goodin, such a position is politically accessible to green politics. However, I am not at all sure that this is correct.

Goodin gets to his somewhat surprising conclusions by starting from a depiction of green values which is staunchly within the tradition which regards nature as a resource. He considers just three possible bases for a political theory of value. These are, first, the instrumental values of commodities in satisfying consumer preferences (the basis of capitalist value theory), second, the labour theory of value (the basis of Marxism and some socialist theories) and finally the green theory. For the last theory, he claims, value is linked to 'some naturally occurring properties of ... objects themselves' (p. 24). As an explication of this, he writes: 'What is crucial in making things valuable ... is the fact that they have a history of having been created by natural processes rather than by artificial human ones' (p. 27). For further explanation, he makes three further claims: '(1) People want to see some sense and pattern to their lives. (2) That requires in turn that their lives be set in some larger context. (3) The products of natural processes, untouched as they are by human hands, provides precisely that desired context' (p. 37).

I do not think Goodin has in fact given three theories of value. The appeal to a larger context and to the independent value of the products of natural processes looks simply to be a further case of the first theory of value. The existence of an appropriately large context is something we might desire for the sake of having meaningful lives. So if meaningful lives are worthwhile things in themselves, the existence of a larger natural context need be no more than instrumentally valuable in bringing this about. Goodin admits this objection as being partially correct. Indeed, in a revealing sentence, he writes: 'Both the neo-classical welfare economic theory of value and the green theory of value as I have presented it link the value of nature to the satisfaction consumers derive from it one way or another' (p. 42). Note the use of the word 'consumers' here. Goodin's analysis is firmly in the tradition which treats nature as a bundle of goods and services, to be used, priced and traded according to human interests. There is no ethical novelty here, and no challenge to conventional understandings of our moral situation in respect of nature.

Economism as an environmental philosophy

Goodin's approach to the question of value in nature is not much different from that taken by a number of British cost-benefit theorists like David Pearce. In trying to give an account of sustainability, Pearce and others have argued that we can at least give some recognition to the value of natural objects and systems by finding means to put economic values on them (see e.g. Pearce *et al.*, 1989, 1990). My own view is that much of this work is morally irrelevant, and that some of the claims on behalf of economic approaches are overblown. To say this is not to deny that economic measures have their uses. However, ethical decisions do not – and indeed *cannot* – reduce to economic ones, and economic methods themselves lack the objectivity which is often claimed for them. Those who take the economist approach to environmental policy are – so I argue – representative of the reductive third trend identified in the opening section. Although there is no space for a detailed demonstration of this, the following discussion will indicate some of the worries that I have about the scope of *economism* as an environmental philosophy.

Let us start by thinking about a case where economic rationality makes sense. Suppose we want to reduce the carbon and sulphur pollution associated with a range of industries. Suppose further that we want to allocate a level of taxation on these pollutants which will be fair. We can approach this problem by considering first of all just how much damage is caused by the pollution and how to put a monetary value on it. Thus for the pollutants mentioned, we can look at effects on health, in terms of treatment costs for patients with respiratory disease directly due to the pollution. We can consider the cost of repairing damaged buildings, the losses suffered by forestry and the loss of agricultural production. All of these costs can be expressed in monetary terms – more or less – and give us a measure of the costs associated with air pollution. Put another way, the same figures indicate the scale of benefits that could be achieved by restriction of such pollution.

Now, it is not a simple matter to fix the appropriate level of sulphur and carbon taxes in the light of the above information. But at least quantifiable costs are associated with pollution in the context described and quantifiable benefits can be gained from controlling it. However, economists typically want to count in other effects of air pollution apart from the ones just mentioned. Consider, for example, the loss of pleasure due to impaired viewing conditions. If air pollution is bad enough in an area with a tourist industry, then there may well be loss of tourist revenue to count in with the other losses. But what of the people who already live or work in the area? Can they not also count in the economic equation? Even if their health is not directly affected, are they not suffering other losses of amenity? And what about trees that start to die, and rivers and lakes which lose their fish stocks through acid deposition? Can these things also be counted in the economic equation?

To put a cost on something which is not already traded is very difficult. Things that do not normally lie inside a set of economic calculations are called 'externalities' and it is interesting to try to find ways of pricing externalities and so include them in a comprehensive calculation. There are many motives for doing this. Think of cheap food, for example, and the farming practices which are involved in its production. In a traditional society in a developing country, around 60% of family income would be spent on food. By contrast, in the industrialized world people spend around 10–20% of family income on food (Thompson, 1994). Such figures may suggest that the industralized countries have very efficient agriculture. But what if this same agriculture causes nitrate pollution of surface and groundwaters, compaction, salinization and loss of soil and leads to future shortages of fresh water? Suppose it also leads to habitat and species loss and the rapid loss of biological diversity in farmed areas. These impacts may not be counted in the farming industry's calculation, or in government figures. So they are externalities. Once we recognize that such impacts are important, we may want to find a way of calculating them, giving them dollar prices and so coming closer to finding 'the true cost' of our food.

The myth of comprehensive valuation

Any talk of 'the true cost' of food, like talk of 'real value', has to be understood as at best metaphorical. The frozen confection I gratefully consume on a hot day may be nutritionally disastrous yet psychologically boosting to my morale. Because I was a long way from the nearest town, the person who sold it to me was able to charge double the usual price for it. Given how hot and tired I was, I would have paid three times the normal price without complaint. So did I get good value for money? What is the real value of that confection? The second question may be a tempting one to ask, but it is misleading. For there is no such thing as 'the real value' of the confection. Buying it was an economic transaction – so it does have an economic value. But, as already indicated, it also has some energy and nutritional value (and disvalue). In fact, it was rather sweeter than I would prefer, so I would not rank it high in aesthetic value. By asking about the 'real value' of the confection, we lure ourselves into thinking that there is some comprehensive scheme which can embrace economics, aesthetics, food value, energy and any other feature of the transaction we care to think about. But to think in this way is mistaken. When we are dealing with issues of more significance than the choice of toothpaste, shampoos and ice-cream, the thought that there is one correct answer to the question of value is potentially disastrous.

Despite this, many economists have expended great energy on thinking up ways to count the cost of everything involved in environmental transactions. David Pearce and his colleagues have tried to lay out systematic ways of 'valuing concern for nature' as they put it (the phrase is the title of Chapter 8 of Turner *et al.*, 1994). Methods of doing this include costing the travel undertaken by people to visit natural areas, valuing environmental factors that impinge on house or land prices and asking people about willingness to pay to preserve or enhance some nature features. But the techniques are not without problems.

Think about questions regarding willingness-to-pay. In one study, values were put on grizzly bears and bighorn sheep by quizzing hunters about how much they were willing to pay to maintain sufficient stocks of these animals (see Chapter 3 of Pearce *et al.*, 1989 for details of this and other studies). This can be useful to know, although it should be obvious that the willingness of hunters to pay for conservation of these animals is only a measure of their 'real value' to hunters. Similar techniques have been widely used in Australia, to find a measure for such things as prefences between eucalypt and pine woodlands, the value of clean air in Sydney and Adelaide, the value Canberra residents place on the existence of the Nadgee Nature Reserve, the tourist value of the Flinders Ranges and the benefits of forest on Fraser Island.

The method starts to become problematic when we push it to extremes. Suppose that all beliefs, values, aesthetic and moral commitments had to be reduced to some common coin – the coin of *informed* or *considered preference.* I use these phrases because they are popular with economists and figure also in discussions outside economics (see, e.g. Attfield and Dell, 1989). If we can measure the 'value' of a forest by seeing how much people are willing to spend to travel to it, then perhaps we can measure the value of freedom by seeing how much people are willing to pay to fight slavery. The attempt to do precisely this is clear in some work done in the 1970s where poverty, slavery and slum housing were argued to be economically inefficient (references in Sagoff, 1988; Chapter 2). Of course, they may be economically inefficient – but that is not what makes them political and social wrongs. If it is recognized that there are beliefs that are not negotiable on markets, that there are values – both moral and aesthetic – that are not to be bought off at any price, and that the rightness and wrongness of actions is unconnected with willingness to pay or receive compensation, then it begins to look as if there may be a great deal of nonsense associated with the attempt to turn all our decisions into economic ones. It would be nothing short of disastrous for moral and political life if the methods for valuing ice-cream and toothpaste as consumer goods are generalized to every aspect of human choice.

I hope that readers have now gathered a hint of what it is at stake in some of the fundamental arguments about environmental ethics and policy. For theorists like Goodin, the status of environmental systems, species and non-human animals appears to be problematic. Although he regards them as providing a large, independent context within which human life makes sense, it seems that he at the same time regards them as akin to consumer goods, to be displayed, consumed or prized according to our informed preferences. For those influenced by the resource economists, it can be hard to understand the challenge posed by environmental ethics as I have described it earlier. Once nature is thought of as a resource, a standing reserve, a set of processes and objects that have the potential to satisfy human desires, it is easy to think of it as a *nothing more* than that. And the challenge posed by any ethic which is properly called 'environmental' is to find a way of articulating the attitude that nature is indeed a great deal more than that.

Despite the moral and political crassness that is a feature of extremist uses of economic methods, especially cost- and risk-benefit analysis, serious theorists are attracted by the method, and it has shown striking resilience to attacks over the last 30 years. In drawing attention to the extremist uses of economic methods, I am not attempting to undermine their proper use in environmental and policy deliberations. Economic issues require economic approaches and treatment. But what I expect general agreement on – without need, I hope, for further argument – is that not all policy issues are purely economic, and certainly that not all questions of value are questions of economic value.

Towards a non-consumerist philosophy of value

If nature is not a warehouse of goods and services from which we as consumers can derive satisfaction, then it might seem that the protection of nature cannot be assimilated to cases such as the protection of ancient monuments, works of art and so on. The approach suggested by Goodin, like that proposed by David Pearce and some other environmental economists, will not be able to get purchase on the problem. But this looks no less disastrous than the extremes of economism. When a policy decision has to be made, we apparently need to have at least two devices available: first, a rational procedure for determining a choice among our options, and, second, the ability to give coherent justifications for the choices we make. By 'coherent' I mean justifications which will make sense to the various interest groups affected by the decision.

One of the attractions of the economic approach is the illusion that cost-benefit and risk-benefit analysis provide the first device. Critics of economic methods have sometimes argued that 'human situational understanding' involves intuition and judgement rather than some counting up of points (Tribe, 1973; Dreyfus, 1982). I think there is some truth in this point when applied both to individual and social decision-making, and I do not share the optimism of Kristin Shrader-Frechette that cost-benefit analysis is a partial substitute for individual wisdom and intuition when a social decision has to be made in a situation where there are thousands (or millions) of opinions to be taken into consideration (Shrader-Frechette, 1984, 1987). However one point on which Shrader-Frechette and I agree is that economic methods provide no help to us in establishing options in the first place. Think back to my example of trying to set a tax on pollutants. Suppose making polluters pay through taxation is one of half-a-dozen options that policy-makers are considering. But why were they only considering half-a-dozen options? At any time, individuals and societies are faced with infinitely many options. We cannot carry out an infinite calculation before settling on our favoured half-a-dozen. So any initial choice of options has to be made without help from economics (Shrader-Frechette, 1987).

Although this point about choice is a simple one, it has very large ramifications which are often overlooked. Some moral theorists have regarded the assessment of consequences of actions as the vital thing in determining whether they are right or wrong. Theories of this sort belong to a genus known as 'consequentialism', the most important species of which is utilitarianism. The fact that we cannot survey the likely consequences of an infinity of actions means that consequentialism in general, and utilitarianism in particular, cannot be adequate moral theories. That is, they cannot give an account of all the factors involved in coming to a moral decision, for the ruling out of infinitely many options cannot itself be explained or justified in consequentialist terms. But, by similar reasoning, economic approaches to social decision-making cannot be complete either. For we have to rule out some options as not worth considering, and focus our economic analysis on only a restricted range of the infinite number available. So the first device is illusory if we think of its scope too generally. At best, we can only use economic methods to choose among options which we regard as (morally, politically, aesthetically, etc.) worth considering.

Economic rationales score much better, it seems to me, when we turn to the second device. They are exactly what 'bureaucratic rationality' requires as a means of justifying decisions that may be unpopular with large numbers of people (Williams, 1985). If we are able to convince ourselves and others that economic approaches do provide a measure of 'the real value' of social and environmental policy options, and if we can devise a suitably comprehensive scheme for such valuing, then we are able to give *principled* reasons for action. The policy-maker can claim to have weighed up all the relevant considerations for and against a course of action; each consideration was in some way commensurable with every other – perhaps because each had a dollar value associated with it – and so it was possible to reach a determinate conclusion, by fair and objective means. The appearance of determinacy, fairness and objectivity is an important tool for those who wield and manipulate power in business, government, and even universities. But this rationalizing use of economic approaches should be resisted by everyone, including economists themselves.

Here is why. If we can persuade ourselves that all values are ultimately no more than consumer preferences, and if everything can have its worth expressed precisely in dollars, then appeal to economic methods will resolve every pressing policy issue. Now everyone knows in their heart that it is lunacy to try to put a price on everything. Remember Oscar Wilde's definition of the cynic as someone who knows the price of everything but the value of nothing. There are some things that it is worse than cynical to put a price on: murder, slavery, rape, kindness, beauty, friendship, love, intimacy, war, oppression, forgiveness, duty – these are just a few of the things whose impact on human lives cannot possibly be measured in dollar terms. It would not be merely cynical to associate dollar costs and benefits with these things – it would be a kind of sickness, a moral sickness that spreads in a society which has forgotten that there are some questions that should not be put, and some answers which should never be given. Once we accept an approach that puts a price on everything we have gone too far. We have entered a trading game and once in that game it is too late to reserve certain options as being beyond price. If there are things that it is wrong to price, then that has to be clear before we start trading in the first place.

Notice what is not being said here. I am not claiming that life insurance is immoral, or that the victims of accidents and violent assault should not be paid compensation. Taking out life insurance is an economic transaction as is the payment under the policy after the insured person's death. But that there are economic aspects to someone's death does not show that death is purely an economic event, or that there is no more value to someone's life than the payout from the available insurance. My point is a pluralistic one. Just as there are economic aspects to most of the things we do, there are also emotional, aesthetic, sensory, political, ethical, intellectual and educational aspects – among others. Each aspect has its own complex of associated values and I can think of no plausible way in which these can be added so as to make a total, comprehensive value. In technical terms, the issue is concerned with what John Rawls called the 'separability of value' (Rawls, 1972). My own view is very strongly opposed to the possibility of aggregating values. Even within a single domain I am a pluralist: e.g., I do not believe that all the moral values in a situation can be aggregated into a single comprehensive moral value, nor that they are comparable in terms of some common standard of value (for more detail on this issue see Brennan, 1992; compare Griffin, 1986; Raz, 1986; Stone, 1988; O'Neill, 1993).

I think it is because of worries like these that those who favour consequentialist and economic approaches try to think of ways to take account of the idea that not every value can be expressed in a common coin (see Attfield and Dell, 1989; and Attfield, 1994, Chapter 15, for an explicit attempt to do this while retaining a 'comprehensive' account of valuing policy options and also maintaining commensurability). The general difficulty of maintaining even weak ways of comparing what is involved in various options while avoiding the pitfall of putting a price on everything is well-described in O'Neill (1993, Chapter 7). To make these points, however, is not to argue that we should dispense with the help that can come from our friends in economics. One approach is to use economic methods (with due care) for aspects of a situation that can be easily priced – e.g., damage to buildings or the medical costs of air pollution. A supplementary approach, as urged by Michael Jacobs and others, is to extend our economic thinking to include an *institutional environmental economics* (Jacobs, 1992, unpublished; and compare Jacobs, 1991). In doing this, we would have to look carefully at the historical particularity (or 'path-dependency') of human institutions and try to develop instruments for analysing these which match, in subtlety and complexity, the attachments, valuations and concerns of the people who work within such institutions.

To come this far is already to have started to build an alternative to the reductive approach urged by consequentialists and some economists. It should be clear by now that a sophisticated approach to policy will not abandon consequentialist and economist thinking, but use it productively for those aspects of planning where it is most helpful. But such an approach will also incorporate very different styles of thinking – drawn from ecology, feminism, the so-called 'deep' ecological movement, the bioregionalist perspective and so on (for a summary introduction to radical environmental philosophies, see List, 1993). Although it may seem rather far-fetched, it seems to me that there are prospects for reconciling aspects of a good institutional economics with some of these positions. In many of them, emphasis is placed on context-dependent features, such as a sense of place, our immediate sense of care and responsibility for humans and other animals in close relationships with us, our attention towards and fondness for small or large features of our local environment, and so on. To the extent that these positions are concerned with the character of human agents, they cannot avoid being concerned ultimately with the character of institutions within which such agents are born, develop, work and forge their identities. As we proceed down this path, it can be expected that Goodin's separation of green values from green agency will look more artificial and less interesting (both ethically and psychologically) than an approach that ties agency and values closely together.

It is interesting to note that cost-benefit theorists like David Pearce often make much of the 'democratic' nature of the economic approach. By sampling consumers, or by measuring travel costs and real estate values, the policy-maker can be sure that thousands or millions of preferences are entered into the great calculation of costs, risks and benefits. Yet the analysis of costs, risks and benefits remains an expert undertaking, requiring expensive studies and carried out only by those with appropriate training and skills. Radical greens, on the other hand, are suspicious of expert assessment and push for democracy in the form of devolved government and the empowering of local people and dispossessed minorities. The devising of appropriate institutional forms of decision-taking thus seems to lead inevitably into questions about our conception of democracy and the forms of government best suited to environmentally-kind society. There is much work to be done on this issue which lies outside the scope of the present paper.

Conclusion

Although I have not dealt in detail with radical environmental philosophies in the present paper, I have tried to give an account of a 'deep' approach to environmental issues. It is common to contrast human-centred and non-human-centred ethics as if there really is a choice to be made here. Likewise, it is tempting to think that we have to choose an economist philosophy or reject it. What I have been arguing, however, is that choosing one perspective rather than another gives us only partial insight into some situations. What I have recommended rejecting is the attempt to reduce all the aspects of environmental choices to economic ones. I would equally strongly reject the reductive move on behalf of aesthetic, consequentialist or any other perspectives if they claimed to give the whole truth of our situation. Reductivism was the third trend identified in the opening section of the paper, and I hope to have at least hinted why it is worth rejecting. To the extent that Goodin, Pearce and Attfield have been tempted by reductive accounts, their analyses are at best incomplete.

We still face a challenge, however. If what is suggested here is correct, then we urgently need to explore the context-dependent development of institutional environmental economics, along with a richer conception of the human moral situation. It is not obvious, for example, how to combine our primary moral concern for other human beings and their welfare with our equally moral concern for other species and the systems that permit their flourishing. To say we need not choose one morality – a human-centred one – over another – a life-centred one, for example – is not yet to tell us how to synthesize the different demands that each moral scheme places on us.

One motive behind the reductive tendencies of those influenced by economics is that we expect to find a system that will organize, explain and even justify our immediate moral responses. The economist systematization may give a rather 'thin' account of our relation to nature, but at least it is systematic. My proposed 'deeper' account seems so far to lack system. This is a serious concern, and it relates to recent pessimistic concerns within mainstream moral philosophy itself. For although we do engage in sophisticated moral judgement and debate as part of everyday living, it is not obvious that there is anything systematic to be found behind all this (Williams, 1985).

Suppose, more optimistically, that a systematic account is to be found. How can we go about finding it? It seems to me that the combination of a sophisticated economism and the sensitive exploration of scientifically sound environmental philosophies offers the prospects of bringing together good philosophy, good ecological science and good social science. Combining these, we can engage in productive explorations of our place in nature and our prospects for maintaining a future human society in which it becomes possible for us to live well in nature. The positive suggestions made in this paper will be helpful, I hope, to others who are struggling to make sense of our situation in the world and are looking for philosophies that will guide us to a relationship with our surroundings that is sophisticated in its understandings, but modest in the demands it makes on natural systems.

References

Attfield, R. (1994) *Environmental Philosophy: Principles and Prospects.* Aldershot: Avebury.
Attfield, R. and Dell, K. (1989) *Values, Conflict and the Environment.* Oxford: Ian Ramsey Centre.
Bentham, J. (1789) *The Principles of Morals and Legislation.* Reprinted 1948, New York: Hafner.
Brennan, A. (1988) *Thinking About Nature.* London: Routledge.
Brennan, A. (1992) Moral pluralism and the environment. *Environmental Values* **1**, 15–33.
Common, M.S., Blamey, R.K. and Norton, T.W. (1993) Sustainability and Environmental Valuation, *Environ. Values* **2**, 299–334.
Dobra, P.M. (1986) Cetaceans: a litany of Cain, In *People, Penguins and Plastic Trees* (D. VanDeVeer, and C. Pierce, eds) pp. 127–34. Belmont: Wadsworth.
Dobson,A. (1990) *Green Political Thought.* London: Harper Collins.
Dreyfus, S. (1982) Formal models vs human situational understanding: inherent limitations on the modelling of business expertise. *Technol. People* **1**, 133–65.
Goodin, R. (1992) *Green Political Theory.* London: Polity Press.
Griffin, J. (1986) *Well-Being.* Oxford: Clarendon Press.
Hume, D. (1739) *A Treatise of Human Nature.* Reprinted 1988, Oxford: Clarendon Press.
Jacobs, M. (1991) *The Green Economy.* London: Pluto Press.
Jacobs, M. (1992) The limits to neoclassicism, unpublished.
List, P. (1993) *Radical Environmentalism.* Belmont: Wadsworth.
Mathews, F. (1991) *The Ecological Self.* London: Routledge.
Murdoch, I. (1970) *The Sovereignty of Good.* London: Routledge & Kegan Paul.
Naess, A. (1973) The shallow and the deep, long-range ecology movement: a summary. *Inquiry* **16**, 95–100.
Naess, A. (1989) *Ecology, Community and Lifestyle: Outline of an Ecosophy* (D. Rothernberg, trans.) Cambridge: Cambridge University Press.
Norton, B. (1991) *Toward Unity Among Environmentalists.* New York: Oxford University Press.
Nussbaum, M. (1986) *The Fragility of Goodness.* Cambridge: Cambridge University Press.
O'Neill, J. (1993) *Ecology, Policy and Politics.* London: Routledge.
Pearce, D., Markandya, L. and Barbier, E. (1989). *Blueprint for a Green Economy.* London: Earthscan.
Pearce, D., Barbier, E. and Markandya, L. (1990) *Sustainable Development.* Aldershot: Edward Elgar.
Rawls, J. (1972) *A Theory of Justice.* Cambridge: Harvard University Press.
Raz, J. (1986) *The morality of freedom.* Oxford: Clarendon Press.
Rolston, H. (1988) *Environmental Ethics.* Philadelphia: Temple University Press.
Sagoff, M. (1988) *The Economy of the Earth.* Cambridge: Cambridge University Press.
Sen, A.K. (1982) *Choice, Welfare and Measurement.* Oxford: Blackwell.
Shrader-Frechette, K. (1984) *Science Policy, Ethics and Economic Methodology.* Dordrecht: D. Reidel.
Shrader-Frechette, K. (1987) The real risks of risk-cost-benefit analysis. In *Technology and Responsibility.* (P.T. Durbin, ed.) pp. 343–57. Dordrecht: D. Reidel.
Stone, C.D. (1988) *Earth and Other Ethics.* New York: Harper & Row.
Taylor, P. (1986) *Respect for Nature.* Princeton: Princeton University Press.
Thompson, P.B. (1994) *The Spirit of the Soil.* London: Routledge.
Tribe, L.H. (1973) Technology assessment and the fourth discontinuity. *Southern California Law Review* **46**, 659.
Turner, R.K., Pearce, D. and Bateman, I. (1994) *Environmental Economics.* London: Harvester Wheatsheaf.
Williams, B. (1985) *Ethics and the Limits of Philosophy.* London: Fontana Books.

Zimmerman, M.E. (1983) Toward a Heideggerian *ethos* for radical environmentalism. *Environ. Ethics* **5**, 99–131.

Zimmerman, M.E. (1993) Rethinking the Heidegger-deep ecology relationship. *Environ. Ethics* **15**, 195–224.

3

The use and abuse of ecological concepts in environmental ethics

ALAN HOLLAND

Department of Philosophy, Furness College, Lancaster University, Lancaster LA1 4YG, UK

This paper looks at some of the ways in which environmental philosophers have sought to press ecological concepts into the service of environmental ethics. It seeks to show that although ecology plays a major role in opening our eyes to sources of value in the natural world, we should not necessarily attempt to build our account of nature's value upon the concepts which ecology supplies. No description is going to capture nature's essence; no formula is going to demonstrate its value. We should recognise the natural world as a particular historic individual and relate to it accordingly. This means acknowledging its value in a contingent, conditional and provisional way, and recognizing its value as a precondition of the value of our own lives.

Keywords: ecology; ethical judgement; moral standing; value

Introduction

The aim of this paper is to offer a brief critical sketch of some current entanglements of ethics with ecology. A few remarks on ethics and ecology are followed by a discussion, illustrating this entanglement, of the idea of respect for the 'order' of nature. We then look at examples of 'order' which feature in ecological discourse and which, at the same time, are thought by some environmental ethicists to represent the natural world in ethically significant ways. After a review of some typical grounds of ethical significance to be found within value theory, there follows a critical discussion of the extent to which it is appropriate to ascribe them to ecological subjects. The paper concludes with some suggestions as to the ways in which it is, and is not, legitimate to use ecology to ground the enterprise of environmental ethics.

Ethics

Although there may have been a religious basis in earlier times for respecting nature as the creation, or even the embodiment, of a supernatural being, the idea of there being a distinctively *ethical* basis for respecting nature is somewhat new, and has only recently begun to surface in mainstream ethical theory. The central question of ethics, in a formulation which goes back to the ancient Greek philosophers is: 'How should one live one's life?'. But, despite the general nature of this formulation, the question has usually been construed rather narrowly, as the question of how human beings should relate to one another.

For some people, even among those who now work on the formulation of an environmental ethic, this remains true. For them, humans remain at the centre of ethical concern, not simply as uniquely capable of expressing such concern, but also as uniquely entitled to receive it. What has changed is not the basis of their ethic, but the realisation of how fundamentally we affect one another's lives through our relationship with our environment. This realisation, to which ecology has made no small contribution, is expressed both in relation to our contemporaries – hence the increase in 'third world' concern – and in relation to our descendants – hence current preoccupations with sustainability, which on one favoured interpretation is held to reflect our concern for future generations (Brundtland, 1987).

For others, however, the new environmental challenges have provoked a more radical response. Drawing inspiration from a number of sources, including ecology, some environmental philosophers claim to have discovered, or rediscovered, a domain of ethical consideration existing alongside, or even outside and independently of, the sphere of human-to-human relations. The more modest claim is that the domain of ethical consideration should be extended to include sentient animals, or living things generally. The more ambitious claim is that it should be extended to include non-living items such as soil, water, sand and rock, or even aggregates of living and/or non-living items such as species, communities, ecosystems, and the planet itself. The ethic embodying such claims is sometimes spoken of as an 'ecological' ethic, and constitutes the chief subject of this paper.

Ecological ethics – some preliminary problems

Of the many problems facing the construction of such an ethic, there are three of a quite general kind which should be signalled right away. (In calling them 'problems' I do not mean to suggest that they are insoluble, but only that their solution may be difficult.)

The first is that once we leave the domain of human affairs, it becomes increasingly unclear what 'showing ethical consideration (or respect)' might consist in. It is not immediately obvious, e.g., what counts as showing respect for a periwinkle. As Wittgenstein says of lions, who are after all comparatively close evolutionary cousins: 'if a lion could talk, we could not understand him' (1967, p. 223). He is pointing out that lions partake of a 'form of life' (habits, practices, ways of interacting, communicating, and so forth) not obviously commensurable with that of humans. If this holds true for humans in their attempt to understand the 'form of life' of lions, how much more problematic is the case of the periwinkle. (At the same time, it should be mentioned that we *are* beginning to see studies, of animal species at any rate, which are precisely aimed at gaining insights into these different forms of life; see e.g. Dawkins, 1980. As a result, we are at least beginning to see what might count as showing consideration for hens, pigs and some of the other domesticated animals.)

A second problem is that once we leave the realm of particular individuals, and raise the question of respect or consideration for larger aggregates such as species, ecosystems or even planets, we enter a realm where ethical thinking becomes uncomfortable, precisely because we begin to lose sight of the claims of individuals. The problems are both moral and logical. Morally, we face the unpalatable prospect of seeing the claims of individuals – those of humans and perhaps also of other living creatures – losing out to those of some larger whole which is judged to be of superior moral significance. If this moral difficulty is solved by retaining equality of value between the individual and the larger whole then, logically, problems arise from the fact that these larger units are themselves composed of individuals. So if a wood comprising, say, a thousand trees, is assigned one unit of value, but so also is each tree which makes up the wood, then we have the anomaly that 1000 units = 1 unit (Sylvan, 1985).

The third problem is that ethicists sometimes step too easily from the claim that an item deserves respect to the claim that we should therefore preserve (or conserve) it, without acknowledging that extra considerations enter into claims of the second kind. There are two aspects to this problem. The first is that it is one thing to demonstrate the value of a thing, but quite another to show that respect for its value has to consist in preserving it. Trivially, the value of disposable cups and plates consists precisely in their disposability. More generally, the value of many things is inherently ephemeral – a smile, a sunset, life itself – and the attempt to prolong them may be ridiculous, inappropriate or even tragic. At the very least, what this point suggests is that, sometimes, it is continuing the possibility of recurrence which matters, rather than continuing something in existence. The more important point is that even if respecting a thing does *prima facie* require that we attempt to preserve it (or refrain from doing it damage or harm), it is quite another question whether, in this particular instance, it actually merits preservation (or protection). The point is that any act of protection or preservation is likely to involve sacrificing or forgoing something else, perhaps something of value. Hence, the justification of such an act involves not simply a judgement of value but a judgement of *comparative* value. Among the most important issues faced by those charged with enacting environmental policy are those which involve enforced choices between alternatives which may all be seen as embodying values of various kinds. For this purpose what is needed is not simply an environmental ethic, dealing with judgments of value, but a conservation ethic, dealing with judgments of comparative value. A number of philosophers have developed evaluative systems for making comparisons *between* environmental 'goods' (Taylor, 1986; Callicott, 1989; Attfield, 1991). But so far there has been little work on evaluative systems for comparing environmental with non-environmental goods; and the gap has tended to be filled by cost-benefit analysis – for want of anything better!

Ecology

The scientific study of nature, and ecology in particular, has helped to stimulate and sustain environmentalism in a number of ways. Chiefly, it has drawn attention to some of the adverse impacts upon the environment which come about in the wake of human economic activity, and to the processes by which these impacts have made themselves felt. It has made people aware, as never before, of the close links that exist between economic and ecological systems. In this way, it has fuelled far-ranging environmental concerns even among those for whom humans remain the centre of ethical attention. But the discipline of ecology has also, through its approach and its ways of conceptualizing the natural world, focused attention upon natural structures and processes which some see as having potential ethical significance in their own right. It has helped to rekindle perceptions of nature akin to those of the 19th-century romantics (Worster, 1977). Whether rightly or wrongly, it has also encouraged some to hope that new models of nature may be emerging to replace the mathematizing and mechanizing paradigms which have prevailed since the times of Galileo and Newton. These latter, in marked contrast to the Aristotelian world view, which presents both organic and inorganic nature as teleological through and through, have proved stony ground on which to attempt to nurture thoughts of value residing in nature itself – except as the expression of some supernatural agency.

Respect for order in nature

A good illustration of the bringing together of scientific description and ethical judgement is provided by a remark of Donald Worster's from his book *Nature's Economy*, where he says that 'One of the most important ethical issues raised anywhere in the past few decades has been whether nature has an order, a pattern that we humans are bound to understand and respect and preserve' (1977, p. ix). Notice in particular the suggestion that there may be certain kinds of order which simply 'command' our respect.

In general, the question whether we are bound to understand and respect nature's 'order' is hardly a recent preoccupation. For many centuries prior to the 'past few decades' it was part of a prevailing western world view to regard nature as *providentially* ordered. In pre-Christian times, the Greek historian Herodotus thought that lions were limited to having only one offspring as a way of tempering their ferocity. And similarly in the Christian era, the 17th-century Englishman Sir Thomas Browne thought that it was providentially arranged for big fierce animals to hibernate so that they would do less mischief, and be less productive (Egerton, 1973). Nor has such an outlook been confined to the western world; many non-western world views saw and still see nature in the same way.

Besides the obligation to understand and respect nature's order, however, Worster also mentions preservation. This is indeed a recent preoccupation whose emergence may be explained in two ways: first, because we can no longer rely on supernatural processes; second, because we can no longer rely on natural ones. In the first place, for as long as a belief in providence prevailed, then although human sin might affect the natural world in ways that people did not *want*, there would never arise any question of the natural world failing to continue in the way that it *should*. A call to preserve what was in any case God's to preserve would have been otiose. In the second place, as we have remarked, the present century has seen a growing realisation of the human capacity to destroy the natural world, and this has given rise to an answering call for measures to be taken for its preservation or protection.

But what, exactly, should we be aiming to preserve? And how convincing is it, after all, to connect respect for the natural world with the perception of it as ordered? These are some preliminary considerations:

(i) The presence of order alone is clearly nowhere near sufficient to provide grounds for respect. For example, the regime of a concentration camp might be exceedingly well ordered but would be unlikely to command respect. The order has to be of a certain kind and, perhaps, to come about in a certain way. For example, there might well be grounds to respect the *providential* order of the natural world out of respect for a providential orderer; but once belief in such an orderer had been abandoned, then the grounds for respect would seem to fall away.

(ii) However, what would remain true, even after belief in such an orderer had been abandoned, is that the natural world would be *such as a providential being might have created*. But would there remain the same grounds for respect – e.g., the same proscription against certain sorts of human meddling with the natural order, once belief in a providential being had been abandoned? This in turn raises the question of whether it is entirely plausible to suppose in the first place that belief in a creator *precedes* admiration for the creation. For, both logically and historically, the direction of conviction has often gone the other way: the character of the creation has been presented as a *reason* for believing in a benevolent creator. This suggests that there may be reasons for respecting the order of the natural world independently of the belief that it is the work of a creator.

(iii) But perhaps certain explanations of how natural order has come about may at least be sufficient to *inhibit* an attitude of respect? We have already noted how an evilly ordered regime would be unlikely to command respect. But the reason for this may lie in the fact that our respect for its order is simply outweighed by our abhorrence at the ends to which it is put. The question is how things stand if we suppose that the order of the natural world has come about purely through natural forces – which may be described as 'blind' at least in the sense that they result from happenstance rather than contrivance. One might test the water a little here by considering, as both analogy and exemplar, the case of the ribbed sand left behind by a retreating tide, which can often display a marvellous symmetry and order. There may be conflicting intuitions about this case: on the one hand, perhaps a reluctance to disturb the sand, but on the other hand a feeling that it can hardly matter if we do, not least because a similar situation is going to be recreated at the very next tide.

(iv) But finally, the question needs to be pressed of how far order is even a necessary factor in any respect we might feel towards the natural world. When Mrs Fanny Turner made the very first bequest to the National Trust (UK) in 1895 of 4 acres of clifflands, she did so in the confidence that 'wild nature' would thereby be able to continue 'having its way'. Such sentiments in regard to nature are hardly unusual – and signal the attraction of the wild and *dis*orderly. At a more mundane level, it is common now to find as a recommended method for planting bulbs throwing them down and planting them where they fall, so as to simulate the randomness of nature. While there may be two different notions of 'wildness' which are getting confused here – 'wild' meaning 'without order' and 'wild' meaning 'without humanly imposed order' – it seems likely enough that it is precisely the (con)fused notion which is the source of appeal. On the other hand, if we were to take seriously Nietzsche's speculation that the world is pure chaos, and that our theories and concepts are mere 'life-preserving errors' (Nietzsche, 1974) it is hard to see how such a world could be the occasion of any kind of respect.

Before probing this line of thought further, however, we shall first look in more detail at some examples of structure and order disclosed by ecological study and the kinds of ethical significance which they are supposed to have.

Ecological structures and processes

It is a central feature of ecology that it deals with assemblages rather than with individuals. It is concerned with questions of distribution and abundance; and only assemblages of various kinds can be spoken of as distributed here and there, and more or less abundantly. Thus ecology deals with animals, vegetables and minerals but it doesn't study individuals of these types as such – which are the province of separate disciplines such as zoology, botany and geology. Its focus, rather, is upon items such as populations, species, habitats, communities and ecosystems. Let us consider three such assemblages with a view to assessing their evaluative potential – species, community and ecosystem. (The discussion is simply illustrative; it has absolutely no pretensions to be comprehensive.)

Species and diversity

That species have an importance over and above the sum of the individual lives of their members is suggested by the different significance attaching to the termination of a life and the termination of a species respectively. Death and extinction are distinct concepts. Death is the end of life, whereas extinction is the end of the opportunity for birth. In the case of human life we see the counterpart of this extra moral dimension recognized in the special opprobrium which is reserved for ethnic cleansing or genocide. Species, therefore, is one category of existent which is thought to be a candidate for ethical recognition in its own right. In addition to species we increasingly find species *diversity* cited as something we should protect, and not simply on grounds of utility. Of course, the functional importance of biodiversity is recognized; but there is sometimes also a recognition that the richness of living forms is a valuable property in its own right – that heterogeneity is somehow better than homogeneity. Part of the case for promoting traditional species-rich meadows over modern pasture dominated by rye grass, for example, rests on the greater diversity of the former.

Prima facie, however, both species and species diversity are problematic as candidates for ethical recognition *in their own right*:

(i) Species are diachronic clusters of individuals linked by lineage. Evolution happens to have favoured such clustering; it might not have done so. Logically, at any rate, it seems possible to imagine individual organisms strung out on a continuum rather than bunched into species. It is not obvious what it is about the species pattern in particular which makes it valuable in its own right. Perhaps it is just as good to have a few very numerous species as many less numerous ones.

(ii) One simple answer is that 'diversity is better'; but diversity is even less obviously a feature to be valued in its own right. Like simplicity, or complexity, it is a qualifying property whose value is at least partly a function of what it qualifies. Simplicity, for example, may be a virtue of theories, but not of jigsaws. The value of species diversity in particular seems to depend upon context: if it is possible to increase the diversity of life-forms in a given habitat, e.g. by the introduction of an 'exotic' form, it does not follow that this is a desirable thing to do.

Community and interdependence

Ecological communities have attracted philosophical interest in two ways, encouragement in both cases coming from ecologists themselves.

(1) Frederic Clements recommended that we view plant communities as organisms. Like organisms, he argued, plant communities exhibit a developmental history beginning with birth and infancy and proceeding through to maturity and senescence, including also the power to reproduce (Clements, 1949). If this were true, communities might qualify for the kind of respect normally reserved for organisms – specifically, respect for their integrity. Clements himself was not so much concerned about the ethical potential of such a view, as with arguing that it promised a predictability which could be useful for conservation projects, and therefore vindicated ecology as a worthwhile enterprise. As an extreme example of this approach, we have the 'Gaia hypothesis' recently propounded by James Lovelock (1979), which invites us to construe the whole planet as a single self-regulating system. (Though here too, rather than promote any ethical message, Lovelock is often at pains to stress the heuristic and predictive value of the hypothesis.)

From its inception Clements's theory met with serious criticism, most notably from Herbert Gleason (1927) and Arthur Tansley (1935). Gleason, appealing to the terminology of mechanics, thought the changes which plant communities undergo, far from being organismic in nature, should be 'compared to a resultant of forces' (1927, p. 302). Both Gleason and Tansley drew attention to the lack of analogy between organisms and plant communities in their formation and development. How far the alleged lack of analogy is due to the choice of animals rather than plants as exemplars of the organism might be questioned, and Tansley concedes that the 'organisation of a mature complex plant association is a very real thing' (1935, p. 291). Indeed, what the organism model does do very clearly is draw attention to the extent of interdependency that there is between the life-forms in a given community, and the extent to which they share a common fate, as do the parts of an organism. By the same token, however, it breaks down to the extent that components of ecological communities live and act for themselves as well as playing a role in a community; and unlike the parts of an organism, many of them are capable of flourishing in communities quite different from those where they may originally be found (Katz, 1985).

(2) Charles Elton hints that we might view animal comunities along the lines of human communities, each species being seen as fulfilling a certain 'role'. When we see a badger, he writes, we should include in our thoughts the idea of its 'place in the community', much as if we should say 'there goes the vicar' (Elton, 1927, p. 64). It is a hint taken up in Leopold's 'land ethic', wherein he urges us to look upon land (i.e. 'soils, waters, plants, and animals') 'as a community to which we belong' (Leopold, 1949). In the human context, at any rate, the community to which we belong can be a source of moral commitment over and above the commitments we have to individual members. If that commitment can be focused on the 'land', we have a basis in ecology for holding that the natural world counts morally (Callicott, 1989).

One challenge to this idea has come from the objection that it trades on an ambiguity (Passmore, 1980). The claim is that a distinction should be drawn between a *moral* community and an *ecological* community. A *moral* community, Passmore argues, is bound together by common interests and mutual obligations – features which are not found in *ecological* communities. Among the questions raised by this criticism are:

(i) whether the proffered characterisation of a moral community is an adequate one and,
(ii) whether it is true that ecological communities lack the features in question.

As regards (i), it might be objected that not all members of a moral community are bound by *mutual* obligations; e.g., babes in arms are members of the moral community without (yet) having any obligations. But Tansley, for one, would uphold the second criticism, being unwilling to 'lump animals and plants together as *members* of a community' in the first place (1935, p. 296). If the term 'community' is not to be too divorced from its common meaning, he argues, then there must be some sort of similarity of nature and status between its members. One might indeed wonder, for example, how far eater and eaten can be spoken of as belonging to the same community. Once again, the ecological item – community – proves a somewhat problematic candidate for ethical recognition in its own right.

Ecosystem, function and health

Philosophers have claimed to discern certain *formal* differences between the concepts of 'ecosystem' and 'community', which are relevant to their potential moral status. In particular:

(i) An ecosystem is defined functionally, a community is defined historically. Roughly, what this means is that a given ecosystem is still in place, if it is continuing to function in a given way, whatever internal changes have taken place. On the other hand, a given community is still in place only if certain internal continuities are preserved. For example, the mechanization of farming can mean that agricultural *systems* are still in place, even while agricultural *communities* are being destroyed. As corollaries we have:
(ii) the components of ecosystems are spoken of as 'parts'; the components of communities are 'members' (although perhaps not exclusively so), and
(iii) the components of ecosystems are thought of as substitutable in a way that the components of communities are not (Katz, 1985). In general:
(iv) the direction and interests of a community tend to be a function of the direction and interests of its members, whereas the parts of an ecosystem are subservient to the functioning of the whole.

These differences have a bearing on the kind of ethical significance, if any, which might attach to ecosystems, as opposed to communities. Unlike our community attachments, the notion of doing things 'on behalf of the ecosystem' does not trip so lightly off the tongue. On the other hand, systems *function* in a particular way, and can function well or badly. Now, given that one of the leading theories of health and disease analyses these concepts in terms of function and malfunction (Reznek, 1987), the way is opened to ascribing these properties not simply to individual organisms but also to ecological structures such as ecosystems. Among the first to moot such an idea was Aldo Leopold, with his notion of 'land-health' which he defines as 'the capacity of the land for self-renewal' (1949, p. 221). The idea has since been taken up in policy circles (Costanza *et al.*, 1992), and has been the subject of a programme of the US Environmental Protection Agency; a journal is being launched devoted to its study. Given the further *normative* force carried by the notions of health and disease – i.e. other things being equal, we ought to promote health and discourage disease – we have here another potential focus for ethical concern.

But once again, doubts accumulate. Whilst it is true that systems can function well or badly, some goal or purpose has to be understood for such terms to get a purchase. A sewage system, e.g., is functioning well if it disposes of sewage efficiently; and we can speak of natural 'systems' functioning well insofar as they serve human purposes. But this only serves to ground instrumental, not ethical value. And despite the best efforts of writers such as Laurence Johnson to prove otherwise (1991), it is quite unclear that natural systems can be ascribed a goal in any other sense (Brennan, 1988). For even if ecosystems could be shown to be 'real', rather than scientific constructions, what Darwin is thought to have shown regarding the natural world in general is how it could be *as if* natural processes were goal directed. Further doubts surround the attribution of health to ecosystems, and the policy implications of such attribution. For example, attributions of health require a theory about the norms of development, longevity etc. of a system; but it is difficult to see how these would be established in the case of ecosystems, which would seem to be essentially *singular* phenomena. On the other hand, if we allow such attributions, it is not clear where we should stop. Does the London Underground, or my study, count as a possible subject? In fact, given Leopold's plausible choice of wilderness as affording a paradigm for the understanding of ecosystem health (a natural system, it might be argued, could no more be unhealthy than the metre rule in Paris could fail to measure one metre), it becomes unclear whether the medical apparatus is doing any work. Ecosystem health reduces to naturalness.

Grounds of ethical significance

What are the grounds, in value theory, upon which these various 'ecological' features might be supposed to merit ethical consideration?

We need, first, to distinguish between a narrower and a broader basis for ethical concern about the things we value. To take the narrower basis first: some things have what might be called 'moral standing' (Attfield, 1991), which means that they qualify directly for moral consideration. Currently, it would be a common view to hold that all human beings have moral standing, and, probably, many animals too; they would have it by virtue of being 'sentient' – i.e. subjects of a life and capable of having experiences. One good test by which to judge whether something has moral standing is to ask whether it can be *wronged*. But what should one say about non-sentient animals (if there are such) and plants? While it might be difficult to argue that plants can be wronged, a more plausible case might be made (though it won't be attempted here) for saying that they can, at least, be *harmed*. They are subject to disease, for example, and premature death – both of which circumstances would be *prima facie* grounds for speaking of harm. For present purposes, let us say that a thing has moral standing if it can be either wronged or harmed. This is not to say that it is necessarily wrong to harm it, e.g. if harming it would prevent some greater harm. Nor is it to say that everything which can be wronged or harmed has *equal* moral standing (see, e.g. Anderson, 1993).

It is clear, however, that we value many things without judging that they have moral standing; and it is here that the broader basis for ethical concern comes into view. If someone were to deface a monument or a painting, we should say that it was damaged or spoiled; but it is unclear whether we would regard it as harmed, and we certainly would not say it was wronged. (Admittedly, for the sake of drawing attention to a real distinction, this is somewhat stipulative; we might well catch ourselves saying of something as innocent as a lemon soufflé, 'will it come to any harm if I put it in the fridge?') Now although monuments and paintings do not have moral standing, there are various ways of not using them well; for example, we can spoil, damage, or simply neglect them. If we do this without good reason, or wantonly, we act wrongly; in particular, we may say that wanton damage of beauty is wrong – ethically. In this way the scope for ethical rectitude or misdemeanour in our relations with the natural world is considerably enlarged. For besides the great range of what may be termed aesthetic qualities, we find there also symbolic qualities to value, sources of enrichment, inspiration, mystery, and an infinite capacity to surprise. Thus, defacing a monument (even an ugly one), disturbing a silence, cheapening a vista, removing a fossil – to name but a few – may all lay us open to ethical remonstration.

With these resources from value theory to deploy, environmental philosophers have selected differing grounds of appeal in their attempts to display the natural world as ethically significant in its own right:

(i) Taking, first, the basis which we have called 'moral standing', some – e.g. those with utilitarian leanings – will see moral significance only where sentient life is involved. For them, the natural world makes a moral claim upon us only insofar as it is itself sentient, or in some way affects the lives of sentient creatures. Among the criticisms which they face there is the charge that their notion of value is too limited, and further, that they cannot explain the basis for certain distinctions which we may want to make e.g. why it might be more important to rescue a rare animal than a common one. Others, in contrast, may adopt an extremely broadminded approach to the matter of moral standing. Thus, if their criterion of moral standing includes the capacity to be harmed, and they also construe items such as species and communities as capable of being harmed, and ecosystems as capable of being rendered unhealthy, they will see the natural world as thick with moral claimants. John Rodman (1977) even sees items such as rivers as possible claimants to liberty, and dams, therefore, as ethically questionable. Among the difficulties faced by this approach is a concern that it views the natural world in too anthropomorphic a way, and a concern that it may lead to repugnant moral conclusions. There is a risk, for example, that the 'land' might generate insatiable moral demands, requiring the sacrifice of individual human lives, as has the 'state' in times gone by.

(ii) Quite a different approach is suggested by writers such as Elliot Sober (1986) and John Lawton (1991). They do not so much look for moral standing in the natural world, but rather construe its value as being primarily aesthetic. Lawton, for example, invites us to view the natural world in the same way as we view a mediaeval cathedral or a Monet painting, and value it as such (1991). An advantage in at least including this approach is that it enables one to attach value to features such as diversity, species-richness and so forth, which – being properties rather than entities – are not the kind of thing which could qualify for moral standing. One drawback which some see in this approach, however, is that it does not invest the claims of the natural world upon us with the requisite degree of seriousness. For example, they might feel that if humans literally and knowingly drive a species to extinction, this is not simply a matter for regret, as when some stately home has been gutted by fire; rather, they feel that the species in question has been *wronged*.

(iii) Finally, many environmental philosophers will adopt an eclectic approach, attributing moral standing to some components of the natural world, whilst finding others valuable because they embody ideals of beauty, significance, mystery and so forth. Such an approach will, of course, inherit both some of the advantages and some of the drawbacks of the two undiluted positions.

The indictment

This brief review of the project of environmental ethics has been selective, and a wide range of issues have been left unexplored. But the question I wish to get to is whether this whole way of approaching the matter of values in nature is correctly focused. I wish to suggest that it is not; and cite the following grounds for disquiet:

Perhaps the first sign that something is wrong is the amount of heaving, stretching and hauling that has to go on, in order to represent communities as organismic, or ecosystems as healthy – i.e., in order to assimilate natural structures and processes to the categories of a preconceived value theory, that of conventional ethics. One is inclined to protest that if we value the natural world, it ought to be for what it is, rather than because it is like something else; what the approach fails to capture is the appeal of nature's 'otherness'. (Hence, Rolston, 1990, is on the right lines, I believe, in his insistence that what he calls 'systemic value' – the value attaching to ecosystems – is *like nothing else*.)

In the second place, such an approach makes the value of the natural world too much hostage to the fortunes of a particular science at a particular time. It seems to ignore the fact that scientific concepts are, or have so far proved to be, eminently disposable, and that the consequence of building ethical judgements upon particular ecological descriptions is likely to be to make these disposable too. Of course we cannot value the natural world without some construction of what it is like; but this is not the same as making the construction *determine* the value. What the cursory historical section (pp. 4–5) has served to suggest is that appreciation of the natural world did not wait upon the emergence of the concept of an ecosystem, or of the 'correct' notion of species, and that sentiments of respect for the natural world survive changes in the way it is described. What is needed, therefore, is some account of respect for the natural world which explains how it can be sustained irrespective (within certain limits) of how the natural world is characterized; and even, how it may have been present *before* the culturally constructed concept of 'nature' arose.

Thirdly, although ecological explanations and descriptions are important, they should not be viewed as specially privileged when it comes to the matter of locating the grounds for valuing the natural world. They are important because, among other things, they help us to predict and manipulate natural events, but it does not follow from this that they have a monopoly on truth. An ecologist may describe mosses and lichens as 'autotrophic'; John Ruskin, on the other hand, refers to them as 'the earth's first mercy' (Ruskin, 1907). It is not easy to say which of these descriptions is 'nearer the truth'. Moreover, insofar as ecological explanations and descriptions serve to help the business of prediction and manipulation, any set of values built upon the concepts so employed will be as likely as not to reflect the values which prompted them (Shrader-Frechette and McCoy, 1994; Howarth, 1995). At the same time, I do not mean to (mis)represent ecology as monolithic and serving only a single set of interests (a misrepresentation well exposed by Strong, 1994). For example, we value the natural world for the way that it challenges us, not only physically but also intellectually; and one of the many other roles which ecologists play is to help to identify these intellectual challenges.

A fourth cause for complaint concerns the exporting from conventional ethics of what might be termed a 'craving' for moral judgements to be demonstrable. In both utilitarian and Kantian ethics, at any rate, it is thought possible to produce *criteria* from which the goodness or badness of states of affairs, and the rightness and wrongness of actions can be deduced. In utilitarianism these criteria derive from the principle of utility – the principle that an action is right insofar as it tends to promote the maximum of happiness; in Kantian ethics, they derive from the categorical imperative – the prescription that one should act only according to a principle that one could will to be a universal law. But it is not for nothing that moral judgements are called 'judgements' – as distinct, say, from 'theorems'. Unlike expressions of preference and feeling, and more so than expressions of opinion or belief, a judgement is open to rational argument; it is something for which we take responsibility, and which we must stand ready to defend. But in the last analysis there can only be a commitment, a leap of faith even; there is no formula to fall back on, or else these would not be circumstances in which *judgement* is called for. (Note that we say we 'can't help' believing something, where it would be odd to confess that we 'can't help' judging something.) Among the mainstream ethical traditions, it is only the tradition going back to Aristotle's ethical writings, specifically his account of practical reasoning, where this seems to be properly acknowledged. Applying the point to environmental ethics, it is far from clear why we should expect the natural world to compel respect in an unconditional and necessary manner. In this regard, some attempts to extort value and meaning from ecological descriptions of the natural world are apt to seem perversely strenuous. And if this is what is meant by the claim that there is intrinsic value in the natural world, namely, that value flows from its very nature (O'Neill, 1992), then that claim has to be questioned.

Finally, there is the question of how far the project of building values upon ecological descriptions is likely to do justice to the appeal of the wild and disorderly. Certainly, many ecologists acknowledge the haphazard qualities of the natural world. Famously, and in opposition to Clements, Herbert Gleason insisted that plant communities were mere 'fortuitous juxtapositions' (quoted in Rolston, 1990, p. 246; *cf.* Gleason, 1927). In a more recent text, John Miles prefers to speak, simply, of 'patches of vegetation' (1979, p. 8). What is interesting is that Rolston's response is to say that if such descriptions were correct, there would be no question of our finding value in plant communities: '[t]here can be no obligations to a fortuitous juxtaposition' (p. 246). But is this true? What might be true is that one would have some difficulty making sense of the idea of valuing something *for being a fortuitous juxtaposition*. But even this is imaginable; the nicely balanced boulder left behind by the retreating ice age can be a great attraction. There is little difficulty, however, about our having grounds for valuing some feature, not *for* being a fortuitous juxtaposition, but which just *happens to be* a fortuitous juxtaposition.

An alternative approach

But all this begs the question: how might we approach things differently? In this final section I shall briefly sketch an alternative approach intended not so much to supplant the mainly 'structural' approaches to value that have been discussed but rather to give them a different, 'historical', focus. (For a rich account of such a perspective, see Cheetham, 1993.)

Part of the strategy of the Roman Stoics to deal with misfortune was to play down the importance of what is particular and personal. If your child or your wife has died, ran the advice (Epictetus, 1977), say to yourself that *a human being* has died, and the loss becomes easier to bear. Usually, and in my view rightly, this approach is thought to miss out something of crucial importance in the sphere of personal relations. First, then, and in a similar vein, I want to suggest that an environmental ethic which concentrates simply on the *kinds* of structures we meet within the natural world, and sees their value as lying simply in the fact that they are things of that kind, is missing out on the crucial importance of the fact that the natural world is a historically particular phenomenon to which we are uniquely related. The subject matter of ecology and the other life sciences does not comprise indefinitely repeatable systems and processes. It is essentially a 'one-way system'. The significance of this fact can hardly be overstated, and is recognized in familiar concerns over irreversibility and extinctions. What I am suggesting is that we have to approach the natural world as a 'this' – as something we can point to but can never exhaustively describe, and that we have to find a way of articulating the value we find there which acknowledges this fact. It is not being suggested that the historical perspective should supersede the structural one, but rather that a recognition of the natural world as a historical individual should be the *controlling* vision, within which the significance of repeatable kinds of order is to be evaluated. It would be a corollary of this view that the life 'sciences', since their subject is a historically particular phenomenon, have more in common with biography or history than many of their practitioners seem likely to want to admit.

At the same time, the feature of being unrepeatable is not peculiar to our natural history, but pertains also to our cultural history. The second point, then, is this: that part of the appeal of the natural world lies in its naturalness. To say this is to endorse positions advocated by Robert Elliot (1982) and Robert Goodin (1992), although not necessarily to endorse their reasons for holding them. Elliot holds that the distinctive value of natural objects lies in their origins – the kinds of processes which have brought them into being. One problem with this view is that he supports the idea of attaching value to origins with reference to the value we attach to original works of art. Certainly this would lead us to regard mimicry of nature as inferior, and therefore to look askance at projects purporting to 'restore' nature. But it is less clear what objection there would be to giving a free hand to a 'Leonardo of landscape' – one who could fashion out of the natural world landscapes as stunning as Mona Lisas fashioned from oils. Goodin therefore is right to look further in attempting to explain the appeal of the natural. The point of the natural world, for Goodin (1992), is the way it sets our lives in a larger context, thereby conferring sense and pattern. The problem with this is that we have the whole universe to go at, so to speak, if we need such a larger context; it is not clear, therefore, why we should not do what we like with the little bit of it which happens to lie close to hand.

A further problem with both views is that they would seem to commit us to attaching the relevant kind of value to the natural world *whatever it was like.* But it seems that the natural world might have been quite different from what it is, and that how it actually is, or indeed any other way that it might have been, is massively improbable. A corollary is that human beings, or creatures exactly like human beings, might have inhabited a natural world which they were not prepared to value, because they did not regard their own lives as worthwhile; there might have been horrible ecosystems, even horribly healthy ecosystems – akin, perhaps, to certain mediaeval depictions of hell. Suppose, for example, a natural world which was a 'hell on earth' for every sentient creature, from which there was no release until it had reproduced its kind. Indeed, the future history of *this* natural world might lead one not to value it; and, of course, a gloomy view of the present natural world is not unknown (Mill, 1874).

However, a condition sufficient for valuing *this* natural world, at least, might be based on a point of Holmes Rolston's. Rolston remarks that 'loving lions and hating jungles is misplaced affection', and that a creature is 'what it is where it is' (Rolston, 1990, p. 258; 249). His point is that one cannot be concerned for lions without also being concerned for the savannahs where they laze – not as an instrumental kind of concern, but because savannahs are partly constitutive of what it is to be a lion. Similarly in the human case, one cannot be concerned for human life, or regard human life as worthwhile, without also being concerned for the (natural) habitats, communities and ecosystems which provide its context. The upshot is that there would be a certain sort of incoherence in valuing human life without also valuing the natural world which has made it possible. On this account, the value attaching to the natural world is conditional, though not instrumental: our valuing the natural world is a condition of our valuing human life, but we do not necessarily value it *because* it is such a condition.

The natural world, then, is a particular historic individual within which human lives are embedded. Other lives also are embedded there which may qualify for moral consideration on the grounds of their 'moral standing'. But otherwise, it has been argued that the natural world has no 'essential' properties, nor any *demonstrable* value. Our valuing the natural world can only be conditional, provisional and contingent. It has to be contingent insofar as the character of the natural world itself is contingent; it has to be provisional insofar as we do not know what the future of this particular world holds in store; it has to be conditional inasmuch as we do not know the present world fully and cannot commit ourselves to value it come what may. However, the natural world envelops and permeates human life in too intimate a way either for its value to us to be construed as purely instrumental, or even, I suspect, for us to adopt the distancing stance of passing judgement (in the sense of a verdict) as to its value. It is more a matter of our relating to it as to an individual, and realising its function as a *precondition* of our own individual existence and our own form of life, insofar as these are things that we value, and of the virtual *absurdity*, therefore, of our standing in judgement upon it.

Revisiting the discussion of 'order' in nature with which we began, it seems appropriate to ask, in view of the account of nature recently given, what kind(s) of order might sensibly be looked for in the narrative of a single historical entity. The logic of the argument suggests that since order is not exclusively synchronic, but also diachronic, which is different, then it is the latter which is of more relevance to the natural world. If ecology is analogous to biography we shall need to consider what are the criteria of an intelligible life-story, and what is to count as (the analogue of) a happy turn of events, a mid-life crisis, a sad ending, and so forth. It seems at least likely that this sort of approach would bear some distinctive fruits in the sphere of conservation objectives and environmental policy generally.

Acknowledgements

I am grateful to the editors of this volume, an anonymous referee, and also to Robin Attfield, John Benson and Anna Holland for their many helpful comments on earlier drafts of this paper.

References

Anderson, J.C. (1993) Species equality and the foundations of moral theory. *Environ. Values* **2**, 347–65.

Attfield, R. (1991) *The Ethics of Environmental Concern, 2nd edn*. Athens, GA: University of Georgia Press.

Brennan, A. (1988) *Thinking About Nature*. London: Routledge.

Brundtland, G.H. (1987) *Our Common Future*. Oxford: Oxford University Press.

Callicott, J.B. (1989) *In Defense of the Land Ethic*. New York: SUNY Press.

Cheetham, T. (1993) The forms of life: complexity, history, and actuality. *Environ. Ethics* **15**, 293–311.

Clements, F. (1949) *Dynamics of Vegetation*. New York: The H.W. Wilson Company.

Costanza, R., Norton, B.G. and Haskell, B.D., eds, (1992) *Ecosystem Health: New Goals for Environmental Management*. Washington, DC: Island Press.

Dawkins, M. (1980) *The Science of Animal Welfare*. London: Chapman & Hall.

Egerton, F.N. (1973) Changing concepts of the balance of nature. *Quart. Rev. Biol.* **48**, 322–50.

Elliot, R. (1982) Faking nature. *Inquiry* **25**, 81–93.

Elton, C.S. (1927) *Animal Ecology*. London: Methuen.
Epictetus (1977) Enchiridion. In *Classics of Western Philosophy*. (S.M. Cahn, ed.) pp. 225–39. Indianapolis: Hackett.
Gleason, H.A. (1927) Further views on the succession concept. *Ecology* **8**, 299–326.
Goodin, R.E. (1992) *Green Political Theory*. Cambridge: Polity Press.
Howarth, J.M. (1995) The crisis of ecology. *Environ. Values* **4**, 17–30.
Johnson, L.E. (1991) *A Morally Deep World*. Cambridge: Cambridge University Press.
Katz, E. (1985) Organism, community and the substitution problem. *Environ. Ethics* **7**, 241–56.
Lawton, J. (1991) Are species useful? *Oikos* **62**, 3–4.
Leopold, A. (1949) *A Sand County Almanac*. Oxford: Oxford University Press.
Lovelock, J.E. (1979) *Gaia: A New Look at Life on Earth*. Oxford: Oxford University Press.
Miles, J. (1979) *Vegetation Dynamics*. London: Chapman & Hall.
Mill, J.S. (1969) Nature. In *Three Essays on Religion*. pp. 3–65. New York: Greenwood Press.
Nietzsche, F. (1974) *The Gay Science* (W. Kaufmann, translation) New York: Vintage Books.
O'Neill, J. (1992) The varieties of intrinsic value. *The Monist* **75**, 119–37.
Passmore, J. (1980) *Man's Responsibility for Nature, 2nd edn*. London: Duckworth.
Reznek, L. (1987) *The Nature of Disease*. London: Routledge.
Rodman, J. (1977) The liberation of nature? *Inquiry* **20**, 83–145.
Rolston III, H. (1990) Duties to ecosystems. In *Companion to a Sand County Almanac* (J.B. Callicott, ed.). Wisconsin: University of Wisconsin Press.
Ruskin, J. (1907) *Modern painters, Vol. 5*. London: Dent.
Shrader-Frechette, K. and McCoy, E.D. (1994) How the tail wags the dog: how value judgements determine ecological science. *Environ. Values* **3**, 107–20.
Sober, E. (1986) Philosophical problems for environmentalism. In *The Preservation of Species: the Value of Biological Diversity* (B.G. Norton, ed.). Princeton: Princeton University Press.
Strong, D. (1994) Disclosive discourse, ecology, and technology. *Environ. Ethics* **16**, 89–102.
Sylvan, R. (1985) *A critique of deep ecology, No. 12, discussion papers in Environmental Philosophy*. Canberra: Australian National University.
Tansley, A. (1935) The use and abuse of vegetational concepts and terms, *Ecology* **16**, 284–307.
Taylor, P. (1986) *Respect for Nature*. Princeton: Princeton University Press.
Wittgenstein, L. (1967) *Philosophical Investigations, 3rd edn*. Oxford: Blackwell.
Worster, D. (1977) *Nature's Economy: a History of Ecological Ideas*. Cambridge: Cambridge University Press.

4

An extensionist environmental ethic

GARY L. COMSTOCK
Bioethics Program, Iowa State University, 403 Ross Hall, Ames, IA 50011–2063, USA

Environmental ethics consists of a set of competing theories about whether human actions and attitudes to nature are morally right or wrong. Ecocentrists are holists whose theory locates the primary site of value in biological communities or ecosystems and who tend to regard actions interfering with the progress of an ecosystem toward its mature equilibrium state as *prima facie* wrong. I suggest that this form of ecocentrism may be built on a questionable scientific foundation, organismic ecology, and that a better scientific foundation for environmental ethics may be found in individualistic neo-Darwinian population biology. However, the latter approach probably requires a corresponding shift away from ethical holism and toward approaches locating value primarily in individuals. I call such environmental ethicists 'extensionists' and briefly outline an extensionist environmental ethic.

Keywords: ethics; environment; holism; individualism; extensionism

Introduction

Environmental ethics covers a variety of positions in moral philosophy, but the following definition is reasonably noncontroversial; environmental ethics consists of a set of competing theories about whether human actions and attitudes to nature are morally right or wrong. As a philosophical discipline environmental ethics has existed for little more than two decades, decades that have seen passionate debate not only about the practical question of how we ought to relate to the natural world but about the theoretical question of how we ought to justify answers to this question. (I trace the beginning of environmental ethics among professional philosophers to Sylvan, 1973; reprinted in Zimmerman, 1993). With regard to the philosophical justification of judgments about issues in applied environmental ethics, several alternatives to anthropocentrism have emerged, including ecofeminism and individualistic biocentrism. (For ecofeminism, see e.g Warren, 1990; for biocentrism see, e.g. Taylor, 1986). The most influential position in the US presently however seems to be holistic ecocentrism. It is a powerful theory.

Ecocentrism

Ecocentrists have developed Aldo Leopold's land ethic into a clearly articulated philosophical position that brings considered moral judgments, moral principles, and background scientific theories into reflective equilibrium around the core notion that humans should not be free to deal with nature as they please. (Parts of this paragraph and parts of my criticism of ecocentrism in later sections of this paper are reprinted with permission of the editor of *Agriculture and Human Values*. See, Comstock, 1995). Rather we ought to judge our actions by Leopold's famous maxim, that an action is morally 'right when it tends to preserve the integrity, stability, and beauty of the biotic community. It is wrong when it tends otherwise' (Leopold, 1949). Ecocentrism is a multifaceted position with different authors developing and justifying the position in various ways. All of them, however, hold that humans have historically conceived of themselves as the rulers and endpoints of nature and consequently have all but ruined it. We must change our self-image to that of 'members and plain citizens' rather than conquerors and lords of the land, and we must desist from treating soil, water, and wildlife as entities valuable only for the instrumental use we can make of them. The way to effect such a change is to shift our attention away from individuals to biotic communities and develop an entirely new, holistic, ethic. Ecocentrists such as J. Baird Callicott advocate 'a shift in the locus of intrinsic value from individuals (whether individual human beings or individual higher 'lower animals') to terrestrial nature – the ecosystem – as a whole' (Callicott, 1989). If ecocentrism is true, we must learn to recognize biological wholes or communities, such as prairie and pine forest ecosystems, and natural entities and processes, such as the hydrologic cycle, as the primary *loci* of intrinsic value. The proposal is a radical one in view of the historical emphasis on individual organisms, especially individual humans, as the primary or sole *loci* of value.

Callicott (1989) gives the following description of the philosophical grounds we should consult in seeking to justify our answers to questions about issues in applied environmental ethics:

> [The] conceptual elements [of ecocentrism] are a Copernican cosmology, a Darwinian protosociobiological natural history of ethics, Darwinian ties of kinship among all forms of life on Earth, and an Eltonian model of the structure of biocenoses all overlaid on a Humean-Smithian moral psychology. Its logic is that natural selection has endowed human beings with an affective moral response to perceived bonds of kinship and community membership and identity; that today the natural environment, the land, is represented as a community, the biotic community; and that, therefore, an environmental or land ethic is both possible ... and necessary, since human beings have collectively acquired the power to destroy the integrity, diversity, and stability of the environing and supporting economy of nature (p. 83).

To what practical judgments do these theoretical considerations lead? Abstract ethical theories become palpable only when we see the specific practices they recommend, and it will be instructive to note representative ecocentric views about traditional human practices.

(i) Trapping. It would be morally permissible to trap beaver and remove their dams:

> to eliminate siltation in an otherwise free-flowing and clear-running stream (for the sake of the complex community of insects, native fish, heron, osprey, and other avian predators of aquatic life ...) (Callicott, 1989, p. 22).

(ii) Meat eating. It would be morally permissible to eat meat if one were involved with the ecological cycle of producing and preparing the meat, but not if the animals were exploited. Therefore fishing and hunting are allowed provided the target species is not endangered (e.g. lynx, cougar, black rhinoceros). In fact, meat eating:

> ... may be more *ecologically* responsible than a wholly vegetable diet [because a wholly vegetarian population would mean a huge expansion of the human population]. Meat, however,

> purchased at the supermarket, externally packaged and internally laced with petrochemicals, fattened in feed lots, slaughtered impersonally, and, in general, mechanically processed from artificial insemination to microwave roaster, is an affront ... to conscience (Callicott, 1989, p. 35).

(iii) Human exploitation of the planet. It is morally desirable to obtain 'a shrinkage, if at all possible, of the domestic sphere; ... [and] a recrudescence of wilderness and a renaissance of tribal cultural experience' (Callicott, 1989, p. 34).

(iv) Wild animals. It is morally wrong to pursue 'commercial traffic in wildlife, zoos, the slaughter of whales and other marine mammals' because wild animals, unlike domesticated animals ... are 'naturally evolved' (Callicott, 1989, p. 32). Therefore, it is wrong to kill elephants for ivory (Taylor, 1986). However, it may be morally required to hunt and kill animals in areas where such actions are 'necessary to protect the local environment, taken as a whole, from the disintegrating effects of [for example] a cervid population explosion' (Callicott, 1989, p. 21).

(v) Agriculture. It is morally impermissible to clear 'more land and [bring] it into agricultural production with further loss of wildlife habitat and ecological destruction' in order to feed 'former animal slaves [i.e., food animals] ..., but not to butcher them' and not to frustrate 'their "natural" behaviour, their right to copulate freely' (Callicott, 1989, p. 31).

Is ecocentrism the environmental ethic we should accept? Thinkers are split on this issue, one group holding that ecocentrism is the new holistic paradigm we need, the other group holding significant reservations. The second group's reservations are based in part on what appears to be a shift in the science of ecology away from an organismic model stressing stability, equilibrium, and holism, and toward a population biology model stressing change, disequilibrium, and individual adaptations. Ecocentrism rests on the organismic model, but as the science shifts sceptics question whether ecocentrism can survive. Meanwhile, the first group thinks the second group timid and wonders whether they have an environmental ethic at all (Rolston, 1988).

Donald Worster writes that turn-of-the-century founders of ecology thought nature consists of broad regularities, 'great coherences that persist over time, giving the landscape a definition of normality.'

> This act of balancing a sense of historical change with a sense of stability, of finding within the swirl of history a normative state, persisted in ecology well into the 1960s. Eugene Odum was its last great exemplar ... For Odum ecology was the study of the 'structure and function of nature,' a definition that almost left out of the picture Darwinian evolution and all its turmoils. ... The [natural] process was orderly, reasonably directional, and predictable, and it culminated in stability, or what Odum called 'maturity.' ... Ecosystems matured but they did not die. They reached a condition of near immortality (Worster, 1985, p. 5).

But there are scientific problems with the organismic model, and they make for corresponding problems in the ethical theory of ecocentrism. The problems coalesce around the problem of identifying ecosystems and their 'natural' progress toward 'maturity'. In order to issue normative rules that will tell moral agents which actions will interfere with the progress of an ecosystem toward its 'unified steady state' and, therefore, which actions are impermissible, an ecological ethic must describe the natural state toward which an ecosystem is progressing. But are there necessary and sufficient conditions the fulfilment of which insure that an ecosystem has reached a state of maturity? Perhaps not, given the diversity of ways in which ecosystems can be described. Consider, e.g., the different ways of understanding one part of an ecosystem, the animal communities of which it consists. How we identify such an assemblage will vary depending upon whether we focus on where animals live, what they eat, or how they are biologically related (Steverson, 1994). Which of these communities is the natural community with which we ought not to interfere? One of them taken singly? All of them taken together? Some subset? The problem of definition plagues other central environmental classification principles including 'abiotic elements', 'natural processes', 'biomes,' and 'guilds'. There is no unique or privileged way to divide up the biological world and each time we divide it up we do so according to principles that serve our scientific needs, allowing us to answer questions arising from the adoption of this or that perspective. And which perspective gives us the 'real' or 'natural' or 'unified steady state' ecosystem? Without an answer to this question we cannot specify which description of an ecosystem provides the scientific standard by which moral judgments may be justified. Without such a standard we cannot know which actions harm, and which benefit, the ecosystem's progress towards its ends.

The second problem is related. Ecosystems may not be the internally directed wholes, not to mention 'super individuals', some ecocentrists assume them to be (Johnson, 1991). This problem is significant because things that possess intrinsic value must be internally directed. If we cannot attribute internal directedness to a moving object, or group of objects, it is conceptually impossible to make sense of the idea of interfering with that object's progress. Only when a being is directed by its own internal ends (e.g. my being directed by my wishes and desires) may we identify actions that have the effect of frustrating the being's progress toward its ends. If a thing is directed by a force external to it, however, we can identify only actions that have the effect of frustrating the external force's progress toward its end. Both a water ski boat and an automatic dialling telephone machine are highly complex systems that engage in movement, but we should not be tempted to ascribe internal directedness to them. (I acknowledge that we may decide one day to ascribe internal directedness to cutting-edge intelligent computer programs, but this is an interesting exception to the rule). To think of a machine as directed toward some goal of its own is counterintuitive. Machines' goals are determined by outside forces: human agents using the machines as means to their ends. To be capable of being harmed in itself, or to be capable of possessing intrinsic value, something must be internally directed.

The view that ecosystems are, at least loosely, internally directed systems gives intelligibility to Holmes Rolston's metaphoric description of ecosystems: 'ecosystems bind life up into discrete individuals and cast them forth to make a way resourcefully through their environment' (Rolston, 1988, p. 183). This view also explains Rolston's interest in defending direct duties to landscapes, vegetation, and natural processes (see p. 96). As holists recognize, the plausibility of ecocentrism seems to rest with the notion that ecosystems have their own goals, goals that provide standards by which to measure whether human actions will aid or destroy an ecosystem's 'health' (Costanza *et al.*, 1992).

Are ecosystems internally directed? Geographical, terrestrial, hydrologic and chemical processes certainly interact within ecosystems in a systematic way, apparently driving them forward. Eugene Odum believed ecosystems have but two stages, 'developmental' and 'mature' and, as Donald Worster puts it, 'The history of nature was by and large reduced to a movement from one category to the other, after which change normally came to an end' (Worster, 1994, p. 6; cf. Worster, 1990 and Callicott, 1992). At the end was the goal, the mature state at which the particular ecosystem aimed. But Odum and others may have overlooked the importance of the fact that determinative influences on the system's development are clearly *external* to the system: rain, sunshine, electromagnetic forces, the effects of tidal patterns. The fact that ecosystems exhibit a complex internal structure is not a sufficient reason to think that they are teleological systems, because the appearance of internal directedness may be as much a consequence or *byproduct* of external forces as an aim generated by the system itself (Cahen, 1988; see also Prigogine and Stenger's 1984). If the structural complexity of an ecosystem arises from forces outside it, then how can we determine which of our actions are contrary to *its* goals? What if *it* has no goals?

The third problem is related. Through time, a geographical location exhibits many apparently mature equilibrium states none of which can simultaneously exist with any of the others. Gary Varner (1993) gives this example, species populations which exist:

> on a given piece of land for only a limited period of time. This is obvious in the case of the [upper Midwestern prairie community] and the [pioneer community of lodgepole pine in the Rockies], which, if they last for more than a few generations, are maintained by allogenic forces, e.g. fire. In the absence of fire, the prairie will soon be replaced by an oak woodland, and the lodgepoles will soon be replaced by fire and hemlock ... on a large enough time scale, every plant community is transient (p. 14).

But which is the natural state that the science of ecology demands we respect? Or is there one? Perhaps my preference for stands of golden aspen in the Rocky Mountains is no less justified than someone else's preference for stands of green Douglas spruce even though in succession the spruce naturally replace the aspen. Perhaps many preferences for how ecosystems should look are based on taste, historically-conditioned aethetic sensibilities, rather than on moral principles allegedly based in environmental science?

Paul Taylor, Tom Regan, Steve Sapontzis, John O'Neill, Bryan Norton, Lilly-Marlene Russow, Gary Varner, Donald Regan, and Elliott Sober are all environmental philosophers who see weaknesses in the ecocentrist attempt to develop an entirely new 'holistic' ethic. (Callicott has modified his position, too, claiming that his view is not at odds with the animal rights position (1988) and renouncing the view that ecosystems and species are individuals with interests. Thanks to an anonymous referee of this journal for making this point. Callicott's writings may have two stages. The later Callicott seems to disavow the early writings on at least three points, arguing that Leopoldian ecocentrism is not in the end incompatible with animal rights views; that ecocentrism does not in the end stand or fall on the truth of the claim that ecosystems are internally directed; and that ecocentrism does not in the end lead to misanthropy. Despite what the later Callicott may say, the structure of his theory remains substantially unchanged and leads one to wonder whether he can have it both ways.) This second group draws on traditional ethical resources such as utilitarianism, rights-based approaches, and pragmatism to construct their environmental ethic. Traditional theories, of course, are individualistic and anthropocentric; they begin and end with the interests and rights of humans. 'Extensionists', however, do not consider humans the only individuals with intrinsic value or moral rights; they argue for recognizing, for example, the moral rights of at least some non-human fauna and the intrinsic value of at least wild flora. The individualism of the traditional approaches is precisely the reason that ecocentrists turned away from them but it is also the reason that this second group is returning to them. As do holists, individualist environmental philosophers develop their theories in slightly divergent ways. But all seek to extend moral considerability to some non-human animals and, in Taylor's and Varner's case, to plants as well. They all see the most fruitful route for environmental ethics in the path of extending traditional ethical frameworks to account for many contemporary environmental intuitions. Extensionists share the traditional ecocentric values of preservation and conservation but they believe they must make the case for these values on the basis of individualistic neo-Darwinian population biology rather than on the basis of holistic organismic ecology.

Extensionism

In the little space I have remaining, I would like to sketch out the briefest outlines of an extensionist environmental ethical theory. Such a theory might be utilitarian in presuming that all pleasures, and pains, of sentient creatures have intrinsic value, or disvalue, regardless of species; and that when we can act without abrogating the rights of individuals, we ought to act so as to maximize the ratio of fulfilled over frustrated individual interests. (Thanks to another anonymous referee for help in formulating this point). The theory might also be rights-based in presuming that as we attempt to promote the general welfare of all subjects of a life that we must simultaneously respect each one's basic rights. We ought not to approve of actions that will allegedly result in the satisfaction of the greatest number of individual interests if such actions require, e.g., the sacrifice of innocent lives. Tragic harms to the few should not be permitted on the basis that doing so will achieve a large number of small benefits for the many.

Which beings qualify as legitimate rights holders? I have given reasons to think that natural objects, systems and processes such as ecosystems and species, should *not* be considered either to be individuals or to be logically possible bearers of moral rights. The welfare of a community, as Bentham argued, is no more than the sum of the welfares of its individual members (Bentham, 1948). How about individual plants, human artifacts, or animals? I would draw the line, very roughly, at the possession of *desire*. A powerful analysis of moral rights holds that a right to *x* requires at a minimum an interest in *x*, so determining which beings are capable of having interests in the relevant sense is critical. The relevant sense of interest is not welfare-interests, i.e. not 'having things that are good or bad for you', because machines, rocks, and distant dying planets have interests in this sense, but they are not morally considerable. The relevant sense of interest is preference-interest, i.e. 'having conative states and the capacity to pursue the objects of them'. Having desires is a paradigmatic sign that a living individual can *take an interest* in this second sense, and the things in which an individual takes an interest provide us with a standard by which to judge whether our actions harm the individual. Of beings with preference interests we can say not only what things are good or bad for them but what things they consider good or bad for them. Many ethicists, e.g., now argue that we ought to recognize certain moral rights for all animals with desires and they have offered a variety of philosophical justifications for this extension: utilitarian (Singer, 1975), rights-based (Regan, 1983), and pragmatic (Sapontzis, 1987). I do not have space here to provide detailed arguments in defense of the animal rights position (see Comstock, 1992a, b, c). Suffice it to say that the soundness of such arguments are crucial to the environmental ethic I propose.

Suppose that all animals with desires are *prima facie* entitled to certain basic moral rights. They matter morally. The next question is, '*How much* do they matter?'. Donald VanDeVeer identifies a principle he calls 'Interspecific Sensitive Speciesism' and formulates ISS as follows:

> When there is a conflict of interests between an animal and a human being, it is morally permissible, *ceteris parabus*, so to act that an interest of the animal is subordinated for the sake of promoting a *like* interest of a human being (or a more basic one) but one may not subordinate a *basic* interest of an animal for the sake of promoting a *peripheral* human interest (VanDeVeer, 1979, p. 183).

To make the implications of ISS clear, note that ISS preserves our intuitions about the superiority, all other things being equal, of humans over dogs in situations in which we must choose to kill one of them, but it does not countenance the torturing of cats or the hunting of elephants for ivory piano keys. In the second case, the elephant's interest in life is basic while the human's interest in ivory is peripheral. When two unlike interests are being compared, if one individual's interest is basic (i.e. the elephant's interest in existence) while the other individual's interest is peripheral (e.g. the human's interest in a luxury material for a musical instrument), then the difference in their species is morally irrelevant and the basic interest is to be respected.

VanDeVeer observes that some will think ISS 'too strong' because it rules out 'having musk perfume, leather wearing apparel or luggage, fur rugs. . . .' and keeping birds in cages merely because some human owner takes pleasure in watching them. VanDeVeer's own objection is that ISS does not allow consideration of a morally relevant factor in adjudicating conflicts of interest, namely that there is 'enormous diversity among the animals whose basic interest may conflict with some human interest' (p. 183). He writes that ISS requires us to weigh equally the interest in not being killed of an oyster and a monkey, a consequence that seems counterintuitive because of the vast differences between these two species. As VanDeVeer writes, 'It is most tempting to think that while both interests are basic, the interest of the chimpanzee is of greater moral weight than that of the oyster ...' (p. 183). In order to avoid this consequence, VanDeVeer proposes another theory:

> When there is an interspecies conflict of interests between two beings, *A* and *B*, it is morally permissible, *ceteris paribus*:
>
> (a) to sacrifice the interest of *A* to promote a like interest of *B* if *A* lacks significant psychological capacities possessed by *B*,
> (b) to sacrifice a basic interest of *A* to promote a serious interest of *B* if *A* substantially lacks significant psychological capacities possessed by *B*,
> (c) to sacrifice the peripheral interest to promote the more basic interest if the beings are similar with respect to psychological capacity (regardless of who possesses the interests).

VanDeVeer calls the principle 'Two Factor Egalitarianism' (TFE). I think there is much to be said on behalf of it. TFE respects our initial assumption that animal interests count, but it also respects our pre-theoretic conviction that animal interests should not be allowed to count equally in all cases with human interests. TFE provides a way both to bring into ethical deliberations morally relevant similarities between humans and other species such as the presence of conative states in some animals, and TFE provides a way to factor in morally relevant differences between species, such as varying mental capacities, without discriminating against individuals simply on the basis of their species membership. The fact that both dogs and humans suffer, therefore, becomes important on TFE. So does the fact that humans, apparently unlike any other species, have the capacity to understand promises, envision our own deaths, and suffer from the imaginative presentation of the suffering of others. Finally, the principle is also action-guiding; it provides useful, if rough, rules to adjudicate interspecific conflicts.

Several clarifications need to be made regarding the idea of comparing different animals' interests. First, not everything we call an interest is morally relevant. We may say that a land formation such as a cliff has an 'interest', e.g., in not being chipped away, but what we mean is that those humans who value the cliff for its rugged appearance or animals who value it for its instrumentality in capturing prey have an interest in the cliff not changing significantly. The 'interest' attributed to the geological formation is not in itself a consideration that directly constrains moral agents. Agents do not have duties directly to the cliff, although they may be indirectly constrained in their actions regarding the cliff by duties they have to desiring beings the fulfilment of whose preference – and welfare – interests depend on the cliff retaining certain of its features.

Second, not all interests are equally important; an interest in having bananas rather than leaves is not as weighty as an interest in keeping a murderous male of one's own species from attacking one's offspring. Upon reflection, it seems that the number of truly *basic* interests are quite limited. There are basic biological needs, or *physiological* interests (e.g. an interest in obtaining adequate food, water, sleep, physical security, protection from the elements and from personal attack by disease or assailants), and basic psychological needs, or *social* interests (e.g. in the case of humans, an interest in having interactions with other humans and animals). Basic interests cannot be sacrificed for long without either loss of life or loss of general well-being, whereas peripheral interests can be foregone forever without serious consequence to one's well-being.

Third, interests vary not only in degrees of importance *to* individuals possessing them but also *between* individuals. Two individuals may have what appears to be an equal interest in the last banana on a tree containing many edible leaves, but the interests may in fact be unequal. One chimpanzee's interest may outweigh the other's if one is seriously deficient in potassium. It seems that we can rank, within limits, conflicts of interests between individuals just as we can rank conflicts of interests within an individual. In cases of conflict between humans in which only one of two competing interests can be satisfied, it is difficult to think of cases in which a basic interest, corresponding to the welfare-interest of chimpanzees in a nutritious diet, would not outweigh a peripheral interest, corresponding to the preference-interest of chimpanzees in having a variety of (what amounts to) superficial dietary choices.

Basic interests (B) need not always triumph over serious interests (S). Consider a case in which a dying woman in an irreversibly comatose state has a healthy liver and is dying of causes unrelated to her liver. Doctors predict that if her liver is taken from her she will die from complications arising from the loss of the liver at exactly the same time as the other difficulties will take her life without the loss of the organ. A second woman needs a liver transplant and will die if she does not have one. She does not need the organ immediately, but she needs it well before the comatose woman is expected to die. Here the first woman's interest in her liver is basic while the second woman's interest in the liver is serious. The second woman's interest is not basic because it is conceivable that the woman can survive without this particular organ, assuming that an organ from another donor becomes available. Suppose that the guardians of the first woman give informed consent for the transplant operation knowing full well that this will have the direct effect of killing the woman, albeit no sooner than she would otherwise have died. It does not seem unreasonable in such cases to prefer one individual's serious interest to another's basic interest.

The example of the woman dying of causes unrelated to her liver suggests that we cannot automatically decide cases where (B) and (S) conflict in favour of (B). Such conflicts may need to be negotiated, and we may not be able to settle them without knowing more about interests potentially satisfied or thwarted *other* than those in direct conflict. The dismal mental state of the comatose patient and her apparent inability to take an interest in anything whatsoever; the hopeless prognosis for her recovery; the acceptance by her guardians of her condition and their interest in trying to make good come out of a painful situation – all of these interests are morally relevant and seem to argue for preferring the serious to the basic interest. So does the recipient's interest in being relieved of worry related to her condition; her interest in a corresponding improvement in her way of life; the interest of her family and friends in securing her future; and her own interests in continuing her career. All of these considerations suggest that the number of interests to be promoted by preferring the serious to the basic interest are more numerous than the interests to be promoted by deciding things the other way.

To forestall misunderstanding, the comparison of other interests in cases of two competing interests is not undertaken in order to discover which interest is in fact the more basic. The comparison of other morally relevant interests is undertaken in order to reach a decision in a case in which a *decision cannot be reached simply by considering the two interests in question*. When interests at the same level are in conflict, as when two basic interests conflict, we call them equal or like interests, meaning that they are both to be weighted identically. Notice that like interests may in fact be identical, as when in a lifeboat case both the dog and the human have an identical interest in 'continuing to exist'. However, like interests need not be identical and may in fact be rather dissimilar, as when one individual has 'an interest in keeping her liver in order to keep bodily functions going' while another individual has 'an interest in obtaining another's liver in order to exist another thirty years, long enough to finish writing my novel'. These two interests are quite dissimilar, and yet they are sufficiently alike to be called equal. When equal interests conflict, each interest counts, in the vivid image of the utilitarian tradition, for one and only one and, in such cases of conflict, the final decision must be made on grounds other than the grounds of the interests in tension.

Note the incommensurability of some interests. Don might have a well-developed sensitivity to pain but no long-range interests while Dan might have no sensitivity to pain yet a deep desire and ability to compose symphonies. They would both have what VanDeVeer calls significant psychological capacities (SPCs), but the SPCs of Don would be irrelevant to cases in which someone's long-range interests would have to be sacrificed. But Don's SPCs would not be irrelevant to decision cases in which someone had to be caused pain. Dan's SPCs would be relevant in all of the cases where Don's were not. Dan's SPCs would be irrelevant in cases in which pain had to be inflicted but relevant in cases in which long-range interests had to be sacrificed. TFEs reliance on a single capacity (SPC) makes it impossible to focus on the complex texture of morally relevant features of different cases (Attfield, 1983).

I have argued that interspecific conflicts of like interests can arise not only between interests at the same level (e.g. basic and basic), but also between different levels (e.g. basic and serious). I suspect, however, that such conflicts are rare and, in general, basic interests should be preferred to serious or peripheral interests. When the conflict is between humans and animals, an extensionist would recommend following the same procedure. In cases of conflict of unequal interests, the more basic interest should be respected regardless of species. In cases of conflict of equal interests, the final decision must be based on maximizing utility by satisfying interests other than those in conflict.

There are reasons that we tend to value the conservation of some species and not others. However, an environmental ethical theory must try to provide good reasons for public policies regarding conservation, and such reasons, I believe, must focus on the interests of individual members of morally considerable species rather than on the alleged needs and interests of species or ecosystems. My view may seem to entail much weaker conservation policies than many environmentalists desire. I suspect, to the contrary, that it would require even more stringent policies, policies that would have the indirect effect of saving many more species than many environmentalists now envision. It would, e.g., require the conservation of sufficient habitat for every spotted owl in Oregon not because each bird is a member of an endangered species but rather because each bird is a subject of a life and having sufficient habitat is a necessary condition of its pursuing its basic interests. It would require the cessation of all hunting and fishing, and not just the hunting and fishing of endangered species such as the lynx and black rhinoceros. It would prohibit slaughter at a young age of 'food animals'. It would oppose the holding of many wild animals in zoos. And it would probably require the conversion of at least some agricultural land for the purpose of providing habitat essential for the flourishing of existing non-human individual animals. In each of these practical judgments my version of extensionism diverges sharply from ecocentrism and, perhaps surprisingly, entails stronger duties of conservation than does ecocentrism.

References

Attfield, R. (1983) *The Ethics of Environmental Concern.* Oxford: Basil Blackwell.

Bentham, J. (1948) *The Principles of Morals and Legislation.* New York: Hafner Press.

Cahen, H. (1988) Against the moral considerability of ecosystems. *Environ. Ethics* **10**, 195–216.

Callicott, J.B. (1988) Animal liberation and environmental ethics: back together again. *Between the Species* **4**, 163–9.

Callicott, J.B. (1988) *Inquiry* **35**, 183–98.

Callicott, J.B. (1989) *In Defense of the Land Ethic: Essays in Environmental Philosophy.* Albany, NY: SUNY Press.

Callicott, J.B. (1992) Aldo Leopold's metaphor. In *Ecosystem Health: New Goals for Environmental Management* (R. Costanza, B.G. Norton and B.D. Haskell, eds.). Washington DC: Island Press.

Comstock, G. (1988) How not to attack animal rights from an environmental perspective. *Between the Species* **4**, 177–8.

Comstock G. (1992a) Pigs and piety: a theocentric perspective on food animals. *Between the Species* **8**, 121–35.

Comstock, G. (1992b) Should we genetically engineer hogs? *Between the Species* **8**, 196–202.

Comstock, G. (1992c) The moral irrelevance of autonomy. *Between the Species* **8**, 15–27.

Comstock, G. (1995) Do agriculturalists need a new, an ecocentric, ethic? *Agriculture and Human Values*, in press.

Costanza, R., Norton, B.G., and Haskell, B.D. eds. (1992) *Ecosystem Health: New Goals For Environmental Management.* Washington, DC: Island Press.

Frankena, W.K. (1979) Ethics and the environment. In *Ethics and Problems of the 21st Century.* (K.E. Goodpaster and K.M. Sayre, eds.) pp. 3–20. Notre Dame, IN: University of Notre Dame Press.

Hettinger, N. and Throop, B. (1994) Can ecocentric ethics withstand chaos in ecology? Unpublished manuscript, quoted with permission.

Johnson, L.E. (1991) *A Morally Deep World: An Essay on Moral Significance and Environmental Ethics*. Cambridge: Cambridge University Press.

Leopold, A. (1966) *A Sand County Almanac With Essays on Conservation from Round River*. New York: Ballantine. (Originally published by Oxford University Press in 1949).

Norton, B.G. (1986) *The Preservation of Species: The Value of Biological Diversity*. Princeton: Princeton University Press.

Norton, B.G. (1991) *Toward Unity Among Environmentalists*. New York: Oxford University Press.

O'Neill, J. (1992) The varieties of intrinsic value. *The Monist* **75**, 119–37.

Prigogine, I. and Stengers, J. (1984) *Order Out of Chaos: Man's New Dialogue With Nature*. New York: Bantam.

Regan, D.H. (1986) Duties of preservation. In *The Preservation of Species: The Value of Biological Diversity* (B.G. Norton, ed.) pp. 195–220. Princeton: Princeton University Press.

Regan, T. (1983) *The Case for Animal Rights*. Berkeley: University of California Press.

Rolston, H. (1988) *Environmental Ethics: Duties to and Values in the Natural World*. Philadelphia: Temple University Press.

Russow, L.M. (1981) Why do species matter? *Environ. Ethics* **3**, 101–12.

Sapontzis, S.F. (1987) *Morals, Reason, and Animals*. Philadelphia: Temple University Press.

Singer, P. (1975) *Animal Liberation*. New York: A New York Review Book, distributed by Random House.

Sober, E. (1986). Philosophical problems for environmentalism. In *The Preservation of Species: The Value of Biological Diversity* (B.G. Norton, ed.) pp. 173–94. Princeton: Princeton University Press.

Steverson, B.K. (1994) Ecocentrism and ecological modeling. *Environ. Ethics* **16**.

Sylvan R. (1973) Is there a need for a new, an environmental ethic? *Proceedings of the XV World congress of philosophy, No. 1*, pp. 205–10. *Varna, Bulgaria.*

Taylor, P. (1986) *Respect For Nature: a Theory of Environmental Ethics*. Princeton: Princeton University Press.

VanDeVeer, D. (1994) Interspecific justice. In *The Environmental Ethics and Policy Book* (D. VanDeVeer and C. Pierce, eds.) Wadsworth: Belmont, CA. (Reprinted from *Inquiry* **22**, VanDeVeer (1979) 55–70).

Varner, G. (1990) Biological functions and biological interests. *Southern J. Phil.* **28**, 251–70.

Varner, G. (1993) A critique of environmental holism. In *In Nature's Interest? Interests, Animal Rights, and Environmental Ethics*. Unpublished manuscript, quoted with permission.

Warren, K.J. (1990) The power and promise of ecological feminism. *Environ. Ethics* **12**, 125–46.

Worster, D. (1985) *Nature's Economy: a History of Ecological Ideas*. Cambridge: Cambridge University Press.

Zimmerman, M.E., ed. (1993) *Environmental Philsophy: From Animal Rights to Radical Ecology*. Englewood Cliffs, NJ; Prentice Hall.

5

Ecology and ethics: relation of religious belief to ecological practice in the Biblical tradition

CALVIN B. DeWITT

Au Sable Institute Outreach Office, 731 State Street, Madison, WI 53711, USA

The Bible, without which Western civilization is inexplicable, has powerful ecological teachings that support an ecological worldview. While these teachings are not widely practised in our time, continuing degradation of ecological systems by humanity requires their re-examination by ecologists and the church. Such re-examination can help develop the mutual understanding necessary for making ethical ecological judgements and putting these teachings into practice in an appropriate manner. Among these teachings are the expectation that people will serve and keep the Creation (earthkeeping principle), that creatures and ecosystems not be relentlessly pressed (sabbath principle), that provisions must be made for the flourishing of the biosphere (fruitfulness principle), that the Earth be filled with biologically diverse and abundant life (fulfilment principle), that pressing the biosphere's absolute limits must be avoided (buffer principle), that people should seek contentment and not selfish gain (contentment principle), that people should seek biospheric integrity rather than self-interest (priority principle) and that people should not fail to act on what they know is right (praxis principle). Ecologists need to recognize and respect these and other biblical ecological teachings and be ready to assist churches in their care and keeping of Creation. And churches must join ecologists in the work of assuring the continued integrity of the biosphere.

Keywords: earthkeeping; sabbath; stewardship; limits; restoration; values

Practical politics

In 1972 I ran for political office in my Town of Dunn – a community just south of the capital city of Madison, Wisconsin. Several citizens decided to come to grips with rampant urbanization of rural land and natural ecosystems in our community of 4000 people. Housing developments had begun to emerge here and there across our 34.5 square miles; agriculture was threatened and so too were our wetlands, lakes and streams. Our decision to replace our local government brought a new Town Board, and a subsequent 2-year moratorium on all land division gave us the peace to conduct an inventory of everything natural and human-made within our borders. Our inventory was extensive, covering the various ecosystems, biodiversity, agriculture, and human community past and present. While it was extensive, it was not complete, and we all came to realise it never could be. Scientific and ecological description provided the data for our implementing a land stewardship plan and codifying it into law in the late 1970s. Despite ongoing skirmishes with those who would destroy this land and its life for immediate personal gain, we now have gained – and hold together – our town with ecological and social integrity. We stand in contrast to communities around us, not merely because of the scientific and ecological knowledge we have gained about our place, but because we decided to act on that knowledge for the benefit of the land and its life. A land ethic has been instilled within us, and we have dedicated our lives to it and its defense. It is an ethic now published in the landscape. It is published in the form of vital and intact ecosystems, restored wetlands, non-structural flood controls, roadsides replanted to prairie, vibrant human community, and much more. Of all Dunn's publications in land and life, perhaps none is more dramatic than its citizens' recent decision to add to its only burial ground – a site unused since the late 1800s. The townspeople have put together the science of their community, they have acted upon it and pursue it with fervour, and they are determined not only to live here but also to be buried in what has become their native place. The people of Dunn have come to know their place and they have come to cherish it.

Max Planck in his essay, *On Religion and Science* (1937), maintained that: 'Man needs science in order to know; religion in order to act'. Do we in the Town of Dunn act religiously? Planck would suggest we do. He observed that:

> our decisions, made by our will, cannot afford to wait until we gain complete knowledge or become omniscient. We stand in the stream of life, surrounded by a multitude of demands and needs. We must often make quick decisions or immediately implement certain plans...

and brought him to say, 'There is no better way to achieve a proper understanding of these remarks than to make the sustained effort to understand more deeply the nature and function of science on the one hand and of religion on the other.'

This, of course, has gotten me to think about the meaning and roles of science, ethics, and religion. Particularly intriguing to me is that Planck saw both religion and science confronting the same struggle – 'a constant, continuing and unrelenting struggle against skepticism and dogmatism, against disbelief and superstition.' What is the role of science and ecological knowledge? What is the role of ethics and of religion?

Planck observed that 'Beliefs about the universe can as little take the place of knowledge and skills as the solution of ethical problems can be achieved through pure intellectual knowledge.' Here he reflects our experience in the Town of Dunn. The Dunn experience and Planck's reference to ethics brings us to an important discovery: our environmental problems are *ethical* problems. Contrary to what our legal and technical approaches to environmental problems may have implied about their sufficiency, legal and technical solutions are not adequate. Laws can and are circumvented and techniques inadequately applied. Without a supportive ethics we may find ourselves looking for 'loopholes to get around the intent of the law', proposing 'mitigations for destructive actions', and developing 'rationales for non-compliance'. Planck and our recent experience help us to come to realize that we are deficient in practical ethics. Meeting this deficiency in practical ethics, with deep belief and fervour, is the behaviour we observe in the Town of Dunn.

Among those who recognized modern environmental problems as ethical problems are ecologists and environmental scientists. Ecologists have frequently treated land degradation and deforestation not merely as things to describe but as situations calling forth an ethical response:

> More than any other single segment of general public today – certainly more than government leaders, lawyers, philosophers, and educators – more, even than most mainline preachers, it is the scientists who are telling us that our world is in critical shape and that the human element is chiefly to blame for it. In fact, there has been a conspicuous about-face in the scientific community within the past two or three decades (Hall, 1986).

Thus, our recent discovery and conviction is this: technical and legal approaches are not sufficient in themselves; they must be joined by ethics. And not only ethics, but ethics put into effective practice. Academic ethics, while beginning to address environmental ethical questions two decades ago, largely has failed, as the philosopher Callicott has pointed out (1991). Confinement of environmental ethics to the academy and to philosophy in particular has done very little practical good. We have found that the academy is not the source or repository of practical environmental ethics. However, religious institutions *are* such, although the modern scientist and citizen may have failed to acknowledge this. As curators of practical ethics, religious institutions may be helpful in addressing environmental issues and problems. The Assisi Declarations (Rinpoche *et al.*, 1986) give us some sense of this through various spokespersons for various religious traditions: 'destruction of the environment and the life depending upon it is a result of ignorance, greed and disregard for the richness of all living things' (Buddhist); we 'repudiate all ill-considered exploitation of nature which threatens to destroy it' (Christian); we should 'declare our determination to halt the present slide towards destruction, to rediscover the ancient tradition of reverence for all life' (Hindu); and 'now, when the whole world is in peril, when the environment is in danger of being poisoned and various species, both plant and animal, are becoming extinct, it is our ... responsibility to put the defence of nature at the very centre of our concern' (Jewish), and people as God's trustees 'are responsible for maintaining the unity of His creation, the integrity of the Earth, its flora and fauna, its wildlife and natural environment' (Muslim).

However stated, the environmental teachings of religion in general, or these religions in particular, while present and even espoused, are not necessarily effective – passively or actively – in bringing appropriate care to the Earth. Chinese-American geographer Yi-Fu Tuan, in his study of environmental situations in several eastern environments 'discovered that, despite their different religious traditions, their *practices* were every bit as destructive of their environments as in the West' (quotation from Livingstone, 1993). David Livingstone points out the significance of Tuan's observation:

> that the 'official' line on [Chinese] attitudes towards environment (the quiescent, adaptive line) in Chinese religions is at odds with what actually is practiced. Quite simply, Chinese mistreatments of nature abound – through deforestation and erosion, rice terracing and urbanization. (1993)

Thus, we can conclude that religion need not necessarily help address environmental problems; we know it may even hinder it. What a given religion or its adherents do depends upon (1) whether belief is put into practice and (2) whether it has provisions for putting belief into practice, and (3) what Huston Smith (1958) refers to as the quality of the religion (Smith, 1958). Beyond these considerations, it is also important, at the very least, to recognize that whatever its contributions, religion is not peripheral to human societies; thus it should not be peripheral in our minds and scholarship.

How might we approach consideration of religion? While many options are possible, I here use an approach that moves directly to the religion and its written text, without which, in the words of Oelschlaeger (1994), Western civilization 'is incomprehensible'. The importance of so doing is illustrated by this philosopher in his book, *Caring for Creation*, in which he describes his 'conversion' from believing religion to be a principle cause of environmental degradation to recognizing religion as a solution. He writes:

> For most of my adult life I believed, as many environmentalists do, that religion was the primary cause of ecological crisis ... I lost that faith by bits and pieces, especially through the demystification of two ecological problems – climate heating and extinction of species – and by

> discovering the roots of my prejudice against religion. That bias grew out of my reading of Lynn White's famous essay blaming Judeo-Christianity for the environmental crisis.

Oelschlaeger continues:

> I think of religion, or more specifically the church – both the public church and congregations of people or fellowships of believers gathered in places of worship, engaging in discourse about their responsibilities to care for creation in the context their traditions of faith – as being more important in the effort to conserve life on earth than all the politicians and experts put together. The church may be, in fact, our last, best chance. My conjecture is this: *There are no solutions for the systemic causes of ecocrisis, at least in democratic societies, apart from religious narrative.*

It is therefore important for us to address religion based upon the Bible, not only because it is the code apart from which Western civilization 'is incomprehensible', but also because it has within it a long-standing Stewardship Tradition that incorporates ecological considerations. The Stewardship Tradition, in the words of Robin Attfield, 'has historically stressed responsibility for nature, and that not only in the interest of human beings...' (Attfield, 1991). He concludes that:

> whatever the causes of the problems may be, our traditions offer resources which may, in refurbished form, allow us to cope with these problems without resorting to the dubious and implausible expedient of introducing a new environmental ethic.

and:

> These traditions ... may well be considered to offer materials from which an environmental ethic equal to our current problems can be elicited, without the need for the introduction of a new ethic to govern our transactions with nature. Indeed, in our existing moral thought and traditions (whether religious or secular) the roots may be found from which, with the help of findings of ecological science, a tenable environmental ethic can grow.

This leads us, of course, to ask what contributions this tradition can make to ecologists and ethical judgements. Analysis of textual material from the Bible provides a set of principles that stand in strong contrast to the eight points presented in Table 1. In Table 2 these biblical principles are presented in parallel with these eight points. Finally, in Table 3 these biblical principles are briefly explained.

The Bible and western civilization

Before considering the Bible, it is important to recognize that there have been many developments and accretions to this code. There are the developments of the Talmud which, through medieval Judaism, produced new interpretations of ecologically based laws and principles in the context of urban society in contrast to the pastoral roots from which it came. There are the developments of the Christian Church that sometimes brought separation of people from the world of nature – from the rest of Creation. And so, while recognizing that many of these derivatives of the Bible were themselves helpful and fruitful elaborations and developments of the Bible itself, I will largely focus on the ecological teachings of the Bible and not its derivatives. I will also allow my probing of the ecological teachings of the texts of the Bible to be informed both by a scientific understanding of the ecological workings of the biosphere and the texts themselves. Reference to land in the Biblical texts will be linked with an ecological understanding of land in the biosphere, and reference to animals and plants in the texts will be connected with an ecological

Table 1. Utilitarian world view

Economy of the biosphere:

1. Earthconsumption – the biosphere and ecosystems are merely resources for consumption.
2. Expendibility/substitutability – ecosystems and species are expendible not only of their surplus but also substance, including extinction.
3. Continuous exploitability – the Earth and the biosphere can be relentlessly pressed without need for restoration or recuperation.
4. Unlimited human population – human population is not limited by resources but only human ingenuity.

Economy of human behaviour:

5. Crisis management – we adjust behaviour only following definitive evidence of substantial consequence.
6. Discontent as best personal condition – people are best motivated by convicting them of discontent irrespective of their situation.
7. Self-interest as best motivation – people are best motivated to serve the needs of society and environment through pursuit of self-interest.
8. Dualism of belief and practice as best policy – people are best when their beliefs are kept separate from their work and actions.

Table 2. Great code world view

Economy of the biosphere:

1. Earthkeeping – serve and keep Creation with all its dynamic integrity and fullness.
2. Fruitfulness (*bal taschit*) – while taking of the fruit of Creation, never destroy its fruitfulness.
3. Restoration/sabbath – provide for adequate time for restoration of the ecosystems you use.
4. Fulfilment and limits – bring yourselves and the creatures under your care to fulfilment, observing the limits.

Economy of human behaviour:

5. Regulation by sabbath – respect and assure the times and places for ecosystem rest and restoration.
6. Contentment – in whatever state you are, learn to be content.
7. Seek system integrity as first priority – seek integrity of Creation first, with justice.
8. Put beliefs into practice – do not fail to act on what you know is right.

Table 3. Utilitarian and great code world views

Economy of the biosphere	
1. Earthkeeping	Earthconsumption
2. Fruitfulness (*bal taschit*)	Expendibility/substitutability
3. Restoration/sabbath	Continuous exploitability
4. Fulfilment and limits	Unlimited human population
Economy of human behaviour:	
5. Regulation by sabbath	Crisis management
6. Contentment	Discontentment as best condition
7. Seek system integrity first	Self-interest as best motivation
8. Put beliefs into practice	Dualism of belief and practice as best

understanding of the biota in their habitats. This approach, of course, is in the spirit of the long-standing 'two-books' tradition, one expression of which was recorded in Belgic Confession of the year 1561 (de Bres, 1959) where under article II entitled, 'The Means Whereby God is Made Known to Us', it is stated:

> We know him by two means:
> First, by the creation, preservation, and government of the universe;
> which is before our eyes
> as a most elegant book,
> wherein all creatures,
> great and small,
> are as so many characters
> leading us to see clearly
> the invisible things of God,
> even his everlasting power
> and divinity,
> as the apostle Paul says (Romans 1:20).
>
> All which things are sufficient to convince men
> and leave them without excuse.
>
> Second, He makes Himself
> more clearly and fully known to us
> by his Holy and divine Word,
> that is to say, as far as is necessary for us
> to know in this life,
> to His glory
> and our salvation.

While our quest here is not for knowledge of God, but rather, for ecological teachings of the Bible, our quest, like that of Article II of the Belgic Confession, is based upon two books: the book of the biosphere and universe, and the Bible. Another traditional support given for this approach is that the Law contained in the Bible is consistent with Law in the natural world and thus they are never in conflict; or, that since these two books are by the same Author, who is just and right, they must be in accord. The problems we find of apparent inconsistency between the two books are our problems, not those of the Author, the text, or the natural world. In using the two-books approach to interpretation, therefore, I am operating on the premise that discovery of environmental ethical principles is possible when both books are read authoritatively and interactively.

A. Economy of the biosphere: four basic ecological principles of the Bible

The teachings of the Bible include four basic principles with great ecological import: the Earthkeeping Principle, the Fruitfulness Principle, the Sabbath Principle, and the Fulfilment and Limits Principle.

1. Earthkeeping Principle: as the Creator keeps and sustains us, so must we keep and sustain the Creator's creation. Genesis 2:15 expects that Adam and Adam's descendants to *serve* and *keep* the garden. The Hewbrew word upon which the translation of *keep* is based is the word 'shamar'.

'Shamar' means a loving, caring, sustaining keeping. This word also is used in the Aaronic blessing, from Numbers 6:24, 'The Lord bless you and *keep* you.' When God's blessing is invoked to keep God's people, it is not merely that God would keep them in a kind of preserved, inactive, uninteresting state. Instead, it is that God would keep them in all their vitality, with all their energy and beauty. The keeping expected of the Creator when the Aaronic blessing is invoked is one that nurtures all of our life-staining and life-fulfiling relationships – with family, spouse, children, neighbours and friends, the land that sustains us, the air and water, and with the Creator. And so too with the keeping of the Garden – in keeping of the Creation. When we *keep* the Creation, we make sure that the creatures under our care and keeping are maintained with all their proper connections – connections with members of the same species, with the many other species with which they interact, with the soil, air and water upon which they depend. The rich and full keeping invoked with the Aaronic blessing is the kind of rich and full keeping that should be brought to the Creator's creatures and to all of Creation.

2. Sabbath Principle: we must provide for Creation's sabbath rests. Exodus 20 and Deuteronomy 5 require that 1 day in 7 be set aside as a day of rest for people and for animals. As human beings and animals are to be given their times of sabbath rest, so also is the land. Exodus 23 commands, 'For six years you shall sow your land and gather in its yield; but the seventh year you shall let it rest and lie fallow, that the poor of your people may eat; and what they leave the wild beasts may eat.' 'You may ask, "What will we eat in the seventh year if we do not plant or harvest our crops?"' God's answer in Leviticus 25 and 26 is: 'I will send you such a blessing in the sixth year that the land will yield enough for three years,' so do not worry, but practice this law so that your land will be *fruitful.* 'If you follow my decrees and are careful to obey my commands, I will send you rain in its season, and the ground will yield its crops and the trees of the field their fruit.' Christ in the New Testament clearly teaches that the sabbath is made for the ones served by it – not the other way around. Thus, the sabbath year is given to protect the land from relentless exploitation, to help the land rejuvenate, to help it get things together again; it is a time of rest and restoration. This sabbath is not merely a legalistic requirement; rather it is a profound principle. And of course, it is not therefore restricted to agriculture but applies to all Creation. The Bible warns in Leviticus 26,

> ... if you will not listen to me and carry out all these commands, and if you reject my decrees and abhor my laws and fail to carry out all my commands and so violate my covenant, ... Your land will be laid waste, and your cities will lie in ruins ... Then the land will enjoy its sabbath years all the time it lies desolate ... then the land will rest and enjoy its sabbaths. All the time that it lies desolate, the land will have the rest it did not have during the sabbaths you lived in it.

3. Fruitfulness Principle: we should enjoy, but must not destroy, Creation's fruitfulness. The fish of the sea and the birds of the air, as well as people, are given the Creator's blessing of fruitfulness. In Genesis 1:20 and 22 God declares, 'Let the water teem with living creatures, and let birds fly above the earth across the expanse of the sky.' And then the Creator blesses these creatures with fruitfulness: 'Be fruitful and increase in number and fill the water in the seas, and let the birds increase on the earth.' Creation reflects the Creator's fruitful work of giving to land and life what satisfies. As it is written in Psalm 104,

> He makes springs pour water into the ravines; it flows between the mountains. They give water to all the beasts of the field; the wild donkeys quench their thirst. The birds of the air nest by the

waters; they sing among its branches. He waters the mountains from his upper chambers; the earth is satisfied by the fruit of his work.

And Psalm 23 describes how the providing Creator '...makes me lie down in green pastures, ... leads me beside quiet waters, ... restores my soul.'

As the Creator's fruitful work brings fruit to Creation, so too should ours. As God provides for the creatures, so should people who were created to image God's care for the world. Imaging God, people too should provide for the creatures. And, as Noah spared no time, expense, or reputation when the creatures were threatened with extinction, neither should we. Deluges – in Noah's time of water, and in our time of floods of people – sprawl over the land, displacing creatures, limiting *their* potential to obey God's command, 'be fruitful and increase in number'.

Thus, while expected to enjoy Creation and expected to partake of Creation's fruit, people may not destroy the *fruitfulness* upon which Creation's fullness depends. While human beings are commanded to increase from their original number (Gen 1:28), so also are fish and birds (Gen 1:22). It is thus clear that human fruitfulness must not be accomplished at the expense of the rest of Creation. Beyond not destroying, we must, with Noah, save the species whose interactions with each other, and with land and water, form the fabric of the biosphere. We should hear the profound admonition of Ezekiel 34:18: Is it not enough for you to feed on the green pastures? Must you also trample them with your feet? Is it not enough for you to drink clear water? Must you muddy it with your feet?

4. Fulfilment and Limits Principle: we should fulfil Creation but within its limits. The fruitful and abundant life, according to the Bible, is not something measured by mere numbers. Achieving biotic potential is not what the abundant life is about. As children are a gift from the Creator to be nourished and treasured, so is the great biotic potential a gift from God – a gift that enables human beings and other creatures to make a rapid come-back following a period of depopulation, or a species in its earlier days to establish itself. Of course if all creatures were to meet their biotic potential life on Earth would be impossible. And this is acknowledged in the Bible in biblical passages on fruitfulness: '...fill the earth, the sky, the seas'. There is no doubt that this 'filling' means a rich, abundant, flourishing fullness – a fullness that envisions Creation so fruitful that it literally swarms with a full spectrum of vibrant and energetic creatures, people included. But it is fruitfulness within bounds and limits. The word for 'fill' used in these texts, and biblical texts on filling wineskins and rivers, means 'filling up' and 'fulfilling'. Both indicate an end point, a capacity, a fulfilment. Thus this blessing is not a blessing without limits; instead, it envisions living creatures, including people, as flourishing within the bounds of God's Law for Creation – within the physical and biological boundaries God has established. Even though the sea is vast and bountiful, and blessed with fruitful and multiplying creatures, limits are set on its size and extent. Even though they are blessed and commanded to fill the seas, the sea creatures are not to overfill them. Thus, the Creator 'made the sand a boundary for the sea, an everlasting barrier it cannot cross' and storks in company with other birds fill and fulfil the skies but do so within God-appointed seasons and spaces. Fruitfulness is commanded, but it has its boundaries.

B. Economy of human behaviour: four basic human behaviour principles of the Bible

5. Sabbath Buffer Principle: we should measure our actions by respect for the sabbaths rather than by meeting absolute limits. How then are we to tell when we as people are at whatever limits there are for the human population? How are we to tell when the force we exercise on the Earth, multiplied by our numbers and our energetic machines, are 'the limit'? The Bible has a remarkable answer in its teachings about the sabbaths. The Bible's answer is to provide a means for sensing limits before they are reached. The sabbath laws say something like this: 'You may go sixth/sevenths of the way to the limit and no further' and 'If you find it necessary to press yourself, your family, your employees, or your animals to work more than six days in seven then you are in danger of transgressing the limits.' Also 'If you find it necessary to stop giving the land time for regeneration by not letting it rest one year in seven, you also are approaching the limits.' The teachings on the sabbath of the week in Exodus 20 and Deuteronomy 5, and those on the sabbath for the land in Exodus 23 and Leviticus 25–26 provide 'buffers' that help us stop short of confronting head-on the crises of species extinctions, starvation, environmental genocide, and the generation of environmental refugees. They state: '... in the seventh year the land is to have a sabbath rest, a sabbath to the Lord', warning that failure to keep these and other ordinances will result in their being driven off the land, and noting that once cleared of its disobedient people that 'the land will rest and enjoy its sabbaths,' 'the rest it did not have during the sabbaths you lived on it'. Observing the sabbaths prevents us from reaching 'the end of our leash'.

Beyond mere law however, the Sabbath laws proclaims a principle: periodic rest, good rulers give the creatures under their care (including themselves) as needed for their rejuvenation, restoration and self-fulfilment. For example, a farmer who lives in the Alberta farming community of Neerlandia gives his land rest every second year, because he practices the sabbath principle from the heart by giving to his land the rest it needs to be fully fruitful to be truly fulfilled. His practice, informed by biblical teachings, comes from the conviction that, 'The sabbath is made for the land and not the land for the sabbath'. There are some places on Earth where once in 2 years is not enough and the land must be in continuous sabbath. The sabbath laws must be practiced from the heart and not just legalistically.

When we find ourselves rationalizing the need to violate sabbath rests and restoration for people, land, and creatures, we are seeking ways to exploit time and space that soon will plunge us and other creatures headlong toward the hard and fearsome limits of ecological collapse, starvation, and extinction.

Thus responsible rule, in the sabbath view, is not to fill every available niche in space and in time with human activity and physical transformation. Instead, life under human rule must be worked significantly short of its absolute limits; Creation must be so ruled that it may enjoy its sabbaths.

Creation must be so nourished with sabbaths that it and all its creatures achieves its God-intended abundant and sustained fruitfulness. The model for all of this is the Creator who on the seventh day, also rested.

> Thus, in the biblical view the Sabbath is more than a buffer zone. The joyful Sabbath rest in the company of God is the telos of creaturely activity – it is the crown of created time (cf. Heschel, 1951).

In the Scriptural view then, the question is not whether we are at the limit, but whether we can and do practice the sabbaths. It is the sabbaths that assure that the ultimate and fearsome limits will not have to be experienced. In the Scriptural view, it also is the question of whether we have been able to achieve our fruitfulness while not diminishing

God's blessing of fruitfulness to the other creatures. For in so limiting ourselves and our use of Creation, the world will be more of what, in the teaching of the Bible, it is intended to be. The whole Creation must have its times and spaces for rest and restoration.

6. Contentment Principle: we must seek contentment as our great gain. The fruitfulness and grace of the Garden – the gifts of Creation – did not satisfy Adam and subsequent generations (Gen 3–11). Even as God promised not to forsake them, they chose to cut out on their own – squeezing ever more from Creation. The Creator wants people to pray: 'Turn my heart to your statutes and not toward selfish gain' (Psalm 119:36). Paul, who has learned the secret of being content (Phil 4:12b), writes: '... godliness with contentment is great gain ...' (1 Tim 6:6–21; also see Heb 13:5).

7. Priority Principle: we must seek first the kingdom, not self-interest. 'This, then is how you should pray: 'Our Father in heaven, hallowed be your name, your kingdom come, your will be done on earth ...' (Matt 6:9–10). It is tempting to follow the example of those who accumulate great gain, to Creation's detriment. But the Bible assures us: 'Trust in the Lord and do good; dwell in the land and enjoy safe pasture ... those who hope in the Lord will inherit the land' (Psalm 37; Matt 5:5). Fulfilment is a *consequence* of seeking the kingdom (Matt 6:33; Zerbe, 1991).

8. Praxis Principle: we must not fail to act on what we know is right. Knowing the requirements of the Bible for stewardship is not enough; they must be practiced, or they do absolutely no good. Hearing, discussing, singing, and contemplating God's message is not enough. The hard saying of scripture is this: We hear from our neighbours, 'Come and hear the message that has come from the Lord.' And they come:

> but they do not put them into practice. With their mouths they express devotion, but their hearts are greedy for unjust gain. Indeed, to them you are nothing more than one who sings love songs with a beautiful voice and plays an instrument well, for they hear your words but do not put them into practice (Ezek 33:30–32; see also Luke 6:46–49).

Believing on God's Son (John 3:16), we must *do* the truth, making God's love for the world plainly evident in our own deeds, energetically engaging in work and action that are in accord, harmony, and fellowship with God, and God's sacrificial love (John 3:21).

Conclusion

The Bible, apart from which Western civilization is inexplicable, has powerful ecological teachings that support an ecological worldview and oppose a utilitarian worldview. This is not to say that these teachings have been widely put into practice in our time – by and large they have not. However, continuing degradation of ecological systems by humanity requires re-examination of these teachings by ecologists and the church. Such re-examination can help develop the mutual understanding necessary for making ethical ecological judgements and putting them into appropriate practice. Ecologists need to recognize and respect biblical ecological teachings and be ready to assist churches in their care and keeping of Creation. And churches must join ecologists in pursuing integrity of the biosphere.

It is not enough to respond or to respond religiously; the response must be of the right kind. Whatever the response, it must be appropriate; it must be in accord with the way ecosystems and the biosphere work. While Huston Smith in his *The Religions of Man* (1958) writes of quality in general terms, in ecological terms quality religion brings people in accord with the principles by which ecosystems work. The result of such quality religion should operate so that ecosystem processes are not degraded or destroyed, and that where systems have been degraded or eliminated that they are restored. This of course implies that the way things are is ecologically right. While this may contradict philosophers such as Immanuel Kant, it still may be right thinking according to environmental philosopher Laura Westra 'Ecological concordance with the way the world works is an important measure that can be applied to evaluate the quality of a given religion and its expression in practice.'

References

Attfield, R. (1991) *The Ethics of Environmental Concern*, 2nd edn. Athens: University of Georgia Press.

Callicott, J. (1991) Genesis and John Muir. In *Covenant For A New Creation: Ethics, Religion, and Public Policy*. (Carol S. Robb and Carl J. Casebolt, eds.) Maryknoll, NY: Orbis.

De Bres, G. (1959) [1561]. Belgic Confession. In *The Psalter Hymnal, Centennial Edition*. Grand Rapids, MI: Publication Committee of the Christian Reformed Church.

Hall, D.J. (1986) *Imaging God: Dominion as Stewardship*. Grand Rapids: Eerdmans.

Heschel, A.J. (1951) *The Sabbath: Its Meaning for Modern Man*. New York: Farrar, Struns and Gironx.

Livingstone, D.N. (1993) The historical roots of our ecological crisis: a reassessment (unpublished paper presented to Christianity Today Institute, Chicago, IL) October 1993.

Oelschlaeger, M. (1994) *Caring for Creation*. New Haven: Yale University Press.

Planck, M. (1937) On Religion and Science, Reprinted in translation. In *On Creation and Science*. Jerusalem: Jerusalem Post Press.

Rinpoche, L.N., Serrini, L., Singh, K., Nasseef, A.O. and Hertzberg, A. (1986) *The Assisi Declarations: Messages on Man & Nature from Buddhism, Christianity, Hinduism, Islam & Judaism*. Geneva: World Wildlife Fund.

Smith, H. (1958) *The Religions of Man*. New York: Harper & Row.

Zerbe, G. (1991) The kingdom of God and stewardship of Creation, In *The Environment and The Christian: What Can We Learn From the New Testament*. (C.B. DeWitt, ed.). Grand Rapids, MI: Eerdmans.

6

Christian theological resources for environmental ethics

BRENT WATERS

Omer E. Robbins Chaplain to the University, University of Redlands, Post Office Box 3080, Redlands, CA 92373–0999, USA

This essay examines how selected Christian doctrines may inform contemporary environmental ethics. Particular attention is focused on the doctrines of creation, redemption and eschatology. The ethical models of dominion, stewardship, and co-creatorship are developed based on these theological doctrines. It is also argued that these ethical models must be explicitly rooted in the church's biblical, creedal and theological tradition if they are to influence the values, virtues and practices of the church. The applicability of these models as resources for developing a public environmental ethic is also assessed.

Keywords: theology; environmental ethics; ethical models; doctrinal theology; stewardship

Introduction

Among many environmentalists religious belief is often viewed, at best, as irrelevant in addressing environmental issues or, at worst – particularly in the case of Christianity – as a leading culprit in creating the global environmental crisis. No religion, either Eastern or Western, primitive or modern, has ever prevented environmental degradation, and in some instances religions have aided and abetted the destruction of ecosystems. This disdain for religion reflects 'the largely unexamined position espoused by scores of ecologists, historians, philosophers, poets, nature writers, political activists, and even some theologians who have identified themselves with the ecology movement' (Santmire, 1985).

Two articles which conveniently frame the growth of popular ecological consciousness over the last quarter-century reflect this environmentalist disdain for religion. In his now classic essay, 'The Historical Roots of Our Ecological Crisis' Lynn White, Jr indicts Christianity as the source of humanity's 'unnatural treatment of nature and its sad results' (White, 1983). According to White, Christian theology stripped nature of any sacred status leaving it composed of inanimate objects and ignorant beasts that humans could exploit and manipulate with impunity. When this anthropocentric faith was uniquely joined with modern science and technology an unprecedented destructive power was unleashed. Nor did Christianity's destructive influence wane with modern secularity. Although 'the forms of our thinking and language have largely ceased to be Christian', we nonetheless continue 'to live ... very largely in a context of Christian axioms' (White, 1983). Consequently, in terms of the global environmental crisis, 'Christianity bears a huge burden of guilt' (White, 1983).

Twenty-five years after the publication of White's essay, Wendell Berry, in his article, 'Christianity and the Survival of Creation', notes that Christians are culpable for the environmental crisis because they ignore the key precepts of biblical faith. Our technological age, which originated and was nurtured with Christendom, ignores the theological belief that the earth belongs to God and humans are called to be God's guests and stewards. The attempt to reshape nature in a technological image is 'the most horrid blasphemy' because it throws 'God's gifts into his face, as of no worth beyond that assigned to them by our destruction of them' (Berry, 1993). This blasphemy is morally and spiritually corrupting leading to the 'preposterous assumption that Paradise can be recovered by violence, by assaulting and laying waste the gifts of creation' (Berry, 1993).

White and Berry both conclude that the environmental crisis stems from a spiritual crisis in which humans attempt to master rather than to live in harmony with nature. In the interval between their two articles, environmental ethics has turned to a variety of resources to solve this crisis of the spirit. In addition to secular appeals to pragmatic and enlightened self-interest, there is growing curiosity in pantheism, mysticism, animistic religions, and the peculiar amalgamation labelled 'New Age' as possible ways to guide the world out of its ecological and spiritual malaise.

Yet this quest for spiritual resources has missed an important cue from White and Berry. Despite their harsh indictment of Christianity's complicity in the environmental crisis, both turn to it to provide spiritual resources for rectifying our current predicament. White proposes St Francis as the 'patron saint for ecologists' (White, 1983), and Berry pleads for the recovery of a stewardship ethic. In short, cutting ourselves off from traditional Christian theological resources will not help heal our ecological and spiritual sickness. We cannot 'devise or invent a new ethic' but we must utilize 'principles' which are 'implicit in our moral traditions' (Attfield, 1983).

There have been a number of Christian responses to the environmental crisis. Many of these works, however, have been defensive attempts to exonerate Christianity, and more positive proposals have largely failed to spark secular or ecclesial interest (Birch and Cobb, 1981; Santmire, 1985; Bowman, 1990; Young, 1994). This deficient utilization of theological resources in developing an environmental ethic is due not only to sinful self-interest and political fickleness on the part of Christians, but also a failure to explicitly ground these proposals in the biblical, creedal and doctrinal traditions of the church which shape the values, virtues and practices of Christian communities.

Rather than defending Christianity against its environmentalist critics or proposing an environmental ethic based on Christian principles, the purpose of this essay is to: (1) review selected Christian theological doctrines to examine if they may serve as resources for an environmental ethic; (2) sketch the contours of ethical models suggested by these theological resources; (3) describe the need for holding these ethical models in tension in order that they may be ecologically sustaining rather than destructive; and (4) assess the prospects of these ethical models for informing a larger environmental ethic beyond ecclesial communities.

Theological doctrines

The first doctrine is *creation*. A central Christian belief is that God is the creator of heaven and earth; i.e. physical and spiritual reality is the result of divine will or intent. Furthermore, the original divine creative act is unique as portrayed by the notion of *creatio ex nihilo*. 'It is a creation, brought into existence by God but something distinct from and over against God' (Peters, 1992). Consequently, the world is not sacred but is rather a divine gift or blessing. This theological doctrine, however, does not imply a static view of creation. *Creatio ex nihilo* is complemented by *creatio continua* in which '[c]reation is properly understood as a continuing act of God's will...' (Polkinghorne, 1986; cf. Peters, 1992). As described by various sciences, creation is a domain of continuous change (Peacocke, 1990). Yet, unlike scientific description, a theological portrayal of creation insists that dynamic natural processes are also influenced by divine intentions. Creation has a purpose, namely to be a hospitable environment for living creatures, suggesting that the 'swarms' mentioned in Genesis is 'the biblical word for biodiversity' (Rolston, 1993). Hence, providence may be understood more in terms of a relationship of interaction between the Creator and creation rather than divine interventions into natural or historical processes. Humans play a unique role is this relationship or interaction because they are capable of discerning and responding to divine disclosures regarding the teleological end or purpose of creation. The principal theological task, then is to truthfully discern these providential disclosures, whereas the primary moral challenge is to respond or act in fitting ways (Niebuhr, 1963).

It is this relation among *creatio ex nihilo*, *creatio continua*, and divine purpose that leads to the next doctrine of *redemption*. The Christian tradition claims that the world (particularly humans) are distorted or plagued by a fundamental flaw in need of correction or healing. In more traditional theological language, all creatures live in a fallen and sinful world. Admittedly, the fall is one of the most difficult theological doctrines to maintain in light of modern science. The natural and anthropological sciences do not portray a pristine relationship between humans and nature that was disrupted in the past.

Yet the fall plays too significant a role in Christian faith to be simply jettisoned. It provides a helpful metaphor for understanding our present ecological predicament; that with the emergence of *Homo sapiens* a species evolved with the mental and technological power to extensively disrupt and destroy various ecosystems, thereby frustrating creation's purpose to be a hospitable environment for diverse expressions of life. Acting upon this potential marks a disordering of creation subjecting it, using Paul's imagery (Romans 8:18–25), to futility. Furthermore, moral and sinful creatures are unable to rescue themselves from this plight. For Christians, the incarnation, death and resurrection of Jesus Christ signifies God's redemptive interaction with creation, enabling it to accomplish its purpose. This redemptive act, however, is not confined to human souls or even human history but includes a cosmic reordering of all creation (Beker, 1980).

It is this redemptive hope that draws us to the final doctrine concerning *eschatology* (the study of 'last things'). The Christian tradition affirms that God is an active and redemptive participant in the *creatio continua*. God sustains and interacts with creation that it might fulfil its purpose. The randomness and chance that characterize our understanding and portrayal of natural and evolutionary processes reflect God's providential ordering of creation toward its consummation (Peacocke, 1979, 1990; Polkinghorne, 1989).

There is an obviously strong teleological dimension in Christian eschatology. God is drawing creation into a redeemed future. As a number of theologians have argued, the preeminent sign of this redeemed future is the resurrection of Jesus from the dead (Pannenberg, 1968; O'Donovan, 1986; Moltmann, 1990). In Christ's resurrection we catch a glimpse of creation's destiny, for not only is Jesus 'the first born of all creation, but he is the first among many' (Gunton, 1992). This redemptive and eschatological sign is not confined to humans but includes all of creation because Christ's resurrection is but a prelude to the resurrection of both history and nature (Moltmann, 1990). Christian hope processes of creative change taking due account both of man's and nature's proper needs, with duly assigned priorities for each' (Peacocke, 1979). Although the tasks of co-creatorship are similar to those of stewardship, the difference lies in their respective orientations. The steward is orientated towards the past and present for the purpose of maintaining an imposed order, whereas the co-creator is oriented toward a relatively open future. Without the complementary role of co-creatorship, stewardship may become a deterministic exercise in which human actions play no significantly creative or redemptive role in fashioning the future. Consequently, Christian theological resources do not support an exclusively preservationist or a restorationist ethic. The image of Christian hope is not a restored Eden, but a new creation:

> The redemption of the world . . . does not serve only to put us back in the Garden of Eden where we began. It leads us on to that further destiny . . . so that the outcome of the world's story cannot be a cyclical return to its beginnings, but must fulfil that purpose in the freeing of creation from its 'futility' (O'Donovan, 1986).

As is true with most, if not all, ethical models, dominion, stewardship, and co-creatorship are accompanied by both moral peril and promise. It is towards briefly assessing this peril and promise that we now direct our attention.

Moral assessment

If these three ethical models are not held in tension, they may prove to be ecologically destructive. Dominion can easily degenerate into antropocentrism. Particularly in a secular age, limited dominion granted by God is often transformed into an idolatry of insatiable human appetite. Consequently, ethical decisions often reflect short-term personal, economic, or political benefit rather than fidelity to God's redemptive intentions for all of creation.

Likewise, when stewardship is rooted in anthropocentrism rather than limited dominion it is used to justify the exploitation of nature. Management becomes the driving force rather than a check on how humans use, manipulate, or destroy ecosystems. The stewards become accountable to no one but themselves. Hence, it is difficult to discern the difference between moral and immoral acts in respect to human conduct within natural processes because there is no larger or future standards against which these acts may be judged.

When the moorings of limited dominion and divinely appointed stewardship are removed from co-creatorship, it can become an exercise in destructive fantasy. Creativity itself becomes the supreme value and goal, and there is often no advance moral standard to discern the difference between what is or is not genuinely creative. Co-creatorship often ignores the reality of evil: 'One never reads of co-creation and sin in the same sentence' (Cole-Turner, 1993). Limited dominion and divinely appointed stewardship acknowledge that our moral decisions are often flawed, which should at least inspire caution and humility in our co-creating efforts.

Despite these perils, these ethical models offer some promising features for environmental ethics if a proper tension is maintained among them. Limited human dominion should be exercised from a theocentric and trinitarian rather than anthropocentric perspective. A theocentric perspective judges which 'actions are right . . . in relation to the sustaining, ordering, limiting, and creative power of God' envisions a new creation as a cosmic reordering encompassing a redeemed relationship among God, humanity and nature. Furthermore, this hope in a new creation informs or shapes our present moral values, acts and virtues.

Furthermore, it needs to be stressed that the doctrines of creation, redemption, and eschatology are not construed within a deistic, monistic or pantheistic understanding of God, but within a uniquely Christian trinitarian framework. There is a necessary relational quality of God both in terms of the divine life within the godhead and its relation to creation. These doctrines stress more the loving nature of the triune God which allows the freedom of 'the other', rather than a deity who is wholly transcendent and independent of, or wholly immanent in and dependent upon, 'the other' (LaCugna, 1991; cf. Moltmann, 1981).

With these brief summaries of creation, redemption, and eschatology in mind, the next task is to sketch some rough contours of three ethical models suggested by these doctrines: dominion, stewardship, and co-creatorship.

Ethical models

The first ethical model is limited *dominion*. Because humans have, or are given by God, certain mental, moral and spiritual capacities, they are called by God to play a unique role in helping creation accomplish its purpose to be a hospitable environment for life. Consequently, humans intervene and regulate various natural processes. Some notion of human dominion is needed to resist the temptation of romanticizing nature. 'We properly fear certain features of the natural world; it is not always a "friend" which serves our best interests' (Gustafson, 1981). Gutstafson also insists that such natural things as germs, viruses, diseases, earthquakes, droughts and floods do not always provide hospitable environments for human life (Gustafson, 1981). It is significant to note that the biblical creation story begins in the *garden* rather than the *wilderness* of Eden, for there was always the expectation that humans would have a duty to tend creation. Dominion implies a hierarchy in creation, yet it is not one of humans being separate from or over and against nature but where they are part of nature playing a unique, tending role.

This limited understanding of dominion leads to the second ethical model of *stewardship*. Dominion is not a license to exploit nature in order to satisfy every human want and need. Rather, humans are authorized to be God's faithful stewards to care or tend creation, enabling it to fulfill its purpose. The 'Bible regards it as man's duty to use nature, not to abstain from using it; but that he must use it as a son of God and in obedience to God's will; and that his use or abuse of nature has far-reaching results in the whole structure of the world... (Moule, 1964). The limits upon human dominion imposed by this stewardship ethic are reflected in the Old Testament laws regarding agriculture and land use (Brueggemann, 1977; Lilburne, 1989), and in New Testament injunctions that Christians should conduct themselves in ways that do not frustrate God's redemptive intentions for creation (Moule, 1964). A stewardship ethic is rooted in the Christian hope that humans will not be saved *from* creation but redeemed *with* all of creation.

It is this eschatological dimension that suggests the final ethical model of *co-creatorship*. An understanding of humans as created co-creators is rooted in the Christian theological traditions of *creatio continua* and hope in the new creation (Peacocke, 1979; Hefner, 1993). Humans work with God in the ongoing creation and redemption of the world. Humans 'become that part of God's creation consciously and intelligently co-operating in the

(Gustafson, 1981). In exercising our dominion, we must ask: 'If God's purposes are for the well-being of the whole of "the creation", what is the place of human well-being in relation to the "whole of creation" '? (Gustafson, 1981). Dominion is exercised, then, in accordance with the Creator's intentions for creation rather than satisfying human wants and desires.

Stewardship practised from a theocentric perspective will act in ways that sustain or enhance the overall well-being of creation. For Christians, the prominent paradigms of stewardship are christological and incarnational. Within the life, death and resurrection of Jesus Christ there are images of servanthood and healing that should be emulated in environmental ethics. Dominion and stewardship certainly imply a natural hierarchy, but in Christian theological terms its apex is a servant and healer of creation rather than its master and owner.

When constrained by theocentric dominion and servant stewardship, co-creatorship may become a more reliable source for accomplishing a redemptive hope. Empowered by the Holy Spirit to act as faithful servants and healers in accordance with God's intentions for creation provides criteria for discerning the difference between genuine and unauthentic creativity. The human use of ecosystems should cohere with an understanding of nature as part of our hope in God's cosmic reordering of a redeemed creation. 'The eschatological transformation of the world is neither the mere repetition of the created world nor its negation. It is its fulfilment, its *telos* or end' (O'Donovan, 1986).

If these models are to lend any promise for developing an environmental ethic, one final, though crucial question remains: What are the prospects of these ethical models influencing a larger or more public environmental ethic?

Prospects for environmental ethics

White (1983) and Berry (1993) argued that the environmental crisis is also a spiritual crisis. Although they condemn Christianity for helping us get into this predicament, they also turn to its spiritual resources to help rescue us from our plight. Seemingly, they both assume that there is a sufficient residue of Christianity in post-christian cultures that an appeal to this faith might provide the necessary resources for formulating a fitting or appropriate environmental ethic.

I do not, however, share their optimism. The 'acids of modernity' (Lipmann, 1929) have had a far more corrosive effect than White or Berry imagine (Gunton, 1993). The theological concepts of creation, redemption, and eschatology no longer have an obvious meaning or pertinence. It is hard to imagine how the ethical models of dominion, stewardship, and co-creatorship could provide the basis for contemporary legal, political and economic policies governing the human use of ecosystems.

It is also doubtful if these theological resources and ethical models will have much influence on the church. The various Christian churches have also been deeply influenced and shaped by modernity, and to date there is little evidence that the environmental crisis has had much effect on Christian beliefs, values, virtues or practices. Yet there is a remnant of hope that some Christian communities might incorporate a serious and sustained commitment to dominion, stewardship and co-creatorship exemplified in their worship, fellowship and piety. Perhaps, over time, others will come to see this more excellent way and be converted to adopt more environmentally responsible lifestyles and policies without necessarily accepting their theological and moral underpinnings. Whether there is sufficient time for this to occur is, of course, the urgency of the issue at stake.

Jacques Ellul (1985) asserts that the image has supplanted the word as the principal form of communication in a technological age. If his assertion is true, then it is incumbent upon Christian communities to become visible signs of properly tending God's creation. In obedience to its divine calling the church will, perhaps ironically, be forced to recover its 'sectarian' origins in order to truly engage the world through its faithful witness to the most genuinely catholic and ecumenical issue of our age. Christians must live as a people in community who love and care for a redeemed creation if they are to proclaim and commend a fitting environmental ethic for our age.

References

Attfield, R. (1983) *The Ethics of Environmental Concern.* New York: Columbia University Press.

Beker, J.C. (1980) *Paul the Apostle: the Triumph of God in Life and Thought.* Edinburgh: T&T Clark.

Berry, W. (1993) Christianity and the survival of creation. *Cross Currents* **43**, 149–63.

Birch, C. and Cobb, J.B. Jr (1981) *The Liberation of Life: From the Cell to the Community.* Cambridge: Cambridge University Press.

Bowman, D.C. (1990) *Beyond the Modern Mind: the Spiritual and Ethical Challenge of the Environmental Crisis.* New York: Pilgrim Press.

Brueggemann, W. (1977) *The Land: Place as Gift, Promise, and Challenge in Biblical Faith.* Philadelphia: Fortress Press.

Cole-Turner, R. (1993) *The New Genesis: Theology and the Genetic Revolution.* Louisville: Westminster/John Knox Press.

Ellul, J. (1985) *The Humiliation of the Word.* (J. Main Hanks, trans.) Grand Rapids: William B. Eerdmans Publishing Co.

Gunton, C.E.(1992) *Christ and Creation: The Didsbury Lectures, 1990.* Carlisle: Paternoster Press.

Gunton, C.E.(1993) *The One, The Three, and The Many: God, Creation and The Culture of Modernity, The Bampton Lectures, 1992.* Cambridge: Cambridge University Press.

Gustafson, J.M. (1981) *Ethics From a Theocentric Perspective: Theology and Ethics.* Oxford: Basil Blackwell.

Hefner, P. (1993) *The Human Factor: Evolution, Culture, and Religion.* Minneapolis: Fortress Press.

LaCugna, C.M. (1991) *God For Us: the Trinity and Christian Life.* San Francisco: HarperCollins.

Lilburne, G.R. (1989) *A Sense of Place: A Christian Theology of the Land.* Nashville: Abingdon Press.

Lippmann, W. (1929) *A Preface to Morals.* New York: Macmillan Co.

Moltmann, J. (1981) *The Trinity and The Kingdom: The Doctrine of God.* (Margaret Kohl, trans.) London: SCM Press.

Moltmann, J. (1990) *The Way of Jesus Christ: Christology in Messianic Dimensions.* (Margaret Kohl, trans.) London: SCM Press.

Moule, C.F.D. (1964) *Man and Nature in the New Testament: Some Reflections on Biblical Ecology.* London: Athlone Press.

Niebuhr, H.R. (1963) *The Responsible Self: An Essay in Christian Moral Philosophy.* New York: Harper and Row.

O'Donovan, O. (1986) *Resurrection and Moral Order: an Outline for Evangelical Ethics.* Leicester: Inter-Varsity Press.

Pannenberg, W. (1968) *Jesus – God and Man.* (Lewis L. Wilkins and Duane A. Priebe, trans.) Philadelphia: Westminster Press.

Peacocke, A.R. (1979) *Creation and The World of Science: The Bampton Lectures, 1978.* Oxford: Clarendon Press.

Peacocke, A.R. (1990) *Theology For a Scientific Age: Being and Becoming – Natural and Divine.* Oxford: Basil Blackwell.

Peters, T. (1992) *God – The World's Future: Systematic Theology For a Postmodern Era.* Minneapolis: Fortress Press.

Polkinghorne, J. (1986) *One World: The Interaction of Science and Theology.* London: SPCK.

Polkinghorne, J. (1989) *Science and Providence: God's Interaction with The World.* London: SPCK.

Rolston, H. III (1993) Environmental ethics: some challenges for Christians. *The Annual of the Society of Christian Ethics.* 163–86.

Santmire, H.P. (1985) *The Travail of Nature: The Ambiguous Ecological Promise of Christian Theology.* Minneapolis: Fortress Press.

White, L. Jr (1983) The historical roots of our ecological crisis. In *Philosophy and Technology: Readings in The Philosophical Problems of Technology.* (Carl Mitcham and Robert Mackey, eds). New York: Free Press.

Young, R.A. (1994) *Healing The Earth: a Theocentric Perspective on Environmental Problems and Their Solutions.* Nashville: Broadman and Holman Publishers.

7

Environmental needs and social justice

J.P. BARKHAM

School of Environmental Sciences, University of East Anglia, Norwich, NR4 7TJ, UK

Despite the growth in awareness and practical action to maintain biodiversity, environmental degradation and ecosystem destruction has continued at a high rate over the last 20 years. The roots of this lie in the predominant international economic order, underpinned by lifestyle demands for increased material consumption. Net flow of wealth from South (less developed) to North (more developed) nations has exacerbated a spiral of increased poverty and environmental degradation in the former. Global environmental conservation depends upon a radical change of direction with the principle of *equity* as the starting point. Notwithstanding the importance of continuing to add to local and small-scale conservation achievements, the prospect of radical change happening seems small, despite it being in the long-term self-interest of the North. The concept of equity is, apparently, unacceptable to Northern electorates.

Keywords: equity; human needs; environmental degradation; poverty; personal psychology

Introduction

It is now more than 20 years since the publication of *The Blueprint for Survival* (Goldsmith *et al.*, 1972) and *Limits to Growth* (Meadows *et al.*, 1972). These were the first documents to give a world-wide airing to ideas developed in more specialist ecological literature over a number of years earlier (e.g. Odum, 1969). They had a profound effect. The essence of their message was twofold; first, that a planet with finite boundaries and resources could not sustain unlimited economic growth and development; and second, that there was mounting evidence to suggest that the cost of the current rate of development was increasingly at the expense of the capacity of the planet's life-support systems to sustain themselves.

The message made intuitive sense, even if at the time it was easy to find fault with some of the detail, the scientific evidence, or the simplicity of the modelling. The scientific establishment expressed outrage and incredulity, led by the Editor of *Nature* (1972) in an editorial rather more hysterical than the hysteria it purported to address. However, an idea whose time has come will not be suppressed by the vested interests of the *status quo*. Thus, it is fascinating to note that just 20 years later, in advance of the United Nations Conference on Environment and Development (The Rio Conference) 1992, the scientific establishment in the form of the Royal Society and the American Academy of Sciences made a first ever joint statement saying:

> if current predictions of population growth prove accurate and patterns of human activity on the planet remain unchanged, science and technology may not be able to prevent either irreversible degradation of the environment or continued poverty for much of the world (Royal Society, 1992).

If the scientific community's considered judgement on the basis of factual evidence, together with the intuitive sense of ordinary people, suggest that the human relationship with the planetary environment threatens our future survival, it makes sense to suggest that there is a scale of problem that demands action. The last 20 years have witnessed a sea-change in awareness and attitude. It is possible also to show many remarkable changes which would have been difficult to imagine 20 years ago – such as an international agreement to limit the release of chlorofluorocarbons (CFCs) into the atmosphere (The Montreal Protocol), an international ban on commercial whaling, together with a multitude of protective measures at a national and local scale, particularly in the First World.

Nevertheless, the message now is still essentially the same as in 1972: 'Unsustainable consumption patterns ... place increasingly severe stress on the life-supporting capacities of our planet' (Agenda 21, 1992). The strong impression given is that the national and international action of the last two decades has merely nibbled at the margins of the problem. The major thrust of predominant human attitudes to the environment, as reflected in action, remain unchanged. These are exploitative, unsustainable, and increasingly threaten life-support systems on an ever larger scale. For example, the annual fish catch from the Black Sea, polluted by the effluent of 16 countries, is now about 12.5% of what it was 10 years ago and only about 20% of what were marketable fish species then are being landed now (Pope, 1994). Again, the remaining pristine temperate rain forests of British Columbia are being destroyed (Tickell, 1993) in just the same way as tropical forests, a process which First World countries are traditionally quick to condemn. Further, 70% of Gross National Product (GNP) is delivered by planetary-degrading activities (Sara Parkin in an address to the Wildlife Trusts National Conference, 10 September 1994). It is telling that the countries with the highest diversity of bird, butterfly and flowering plant species, and almost certainly biodiversity in general, also happen to be tropical countries with some of the lowest GNPs.

At the same time, the British Prime Minister is speaking about plans to double standards of living in Britain over the next 25 years. Leaders of other political parties echo similar sentiments. This is indicative of the primacy being given to traditional growth economics. It is no surprise therefore that, in a country whose instruments for environmental protection are arguably as well-developed as any in the North, the destruction of remaining areas of seminatural ecosystems in Britain continues inexorably (Table 1) and at a rate not dissimilar to that of the tropical moist forests of the Amazon Basin and the boreal ones of Canada. The results are not so spectacular but are indicative of fundamentally the same economic process. The difference in scale of destruction of tiny remnants of British seminatural ecosystems and large tropical areas and boreal ones highlights the twin purpose of conserving biodiversity. We have a responsibility to manage the natural world in a way that safeguards not only future human material welfare, but also its cultural and spiritual values through cherishing non-human life.

It appears that, although the achievement of the recent past has been a growing awareness of the problem, the responsive action is still far from the top of the political agenda. There are reasons for this and they have little to do with shortcomings in ecological and scientific knowledge. There are bigger actors in the play. What and whose are the dominant agendas? The purpose of this paper is to explore this question. At the outset it is necessary to identify basic human needs and behaviour.

Table 1. Recent losses of biodiversity in English rural counties due to site destruction from development and agricultural intensification, compared with percentage areas of forest destroyed in the Brazilian Amazon and in Canada

Shropshire:
1978–1988 loss of sites of nature conservation interest 10.3%
Cornwall:
1980–1988 sites of nature conservation interest suffering partial or complete habitat loss 11%
Brazil:
Area of forest destroyed in Amazon basin 12%
Canada:
Area of forest destroyed 10.3%

(Sources: Royal Society for Nature Conservation, 1989; Anon, 1993).

A hierarchy of basic needs

Maslow (1970) identified the fact that human beings have a variety of needs they seek to fulfil, necessarily in a particular order (Table 2). Physiological needs are those required to satisfy basic bodily functions. These come first. Once satisfied, there is a quest for safety. Belongingness and love are obviously required for healthy emotional development and continue through adult life. The higher order needs for self-esteem and self-fulfilment are attended to satisfactorily only when the lower order ones have been met. Such a hierarchy is attended to not only through the span of life as a whole, but also on a daily basis.

With around 1 billion people in the world malnourished or starving, together with populations of a number of regions displaced through war and civil strife, it becomes readily apparent that a sizeable part of the world's population never, or rarely, have the opportunity to meet the higher order needs which pre-occupy most of those living in the First World. As Maslow (1970) goes on to point out, when a person does not know when and from where the next meal is coming from, higher needs are irrelevant:

> Freedom, love, community feeling, respect, philosophy, may all be waved aside as fripperies that are useless, since they fail to fill the stomach. Such a man may fairly be said to live by bread alone.

Table 2. A hierarchy of basic human needs, these are attended to in order from the bottom upwards

Need for self-fulfilment
Esteem needs
Belongingness and love needs
Safety needs
Physiological needs

(Source: Maslow, 1970)

Further, there is a link between a collection of unmet needs in the individual and aggression:

> The child who is insecure, basically thwarted, or threatened in his needs for safety, love, belongingness, and self-esteem is the child who will show more selfishness, hatred, aggression, and destructiveness. (Maslow, 1970)

Unmet infantile needs have consequences. These may not just be lowered ability to cope with the difficulties of adult life and lower effectiveness in society and work, but also a lower tolerance of deprivation:

> It is precisely those individuals in whom a certain need has always been satisfied who are best equipped to tolerate deprivation of that need in the future ... (Maslow, 1970)

When there is a collective of hungry and displaced people suffering from the consequences of deprivation of food, safety, love and belonging, a much wider threat to security and stability becomes apparent. As Enzensberger (1991) has pointed out:

> This does not mean that the enemy of humanity can emerge under any conditions, suddenly and without warning. The pre-condition for him finding followers who yearn for destruction is a long-standing despair of millions of people, a collective humiliation which has corroded their self-respect to the core.
>
> When a collective no longer sees any chance of finding compensations for the humiliations, real and imagined, heaped upon it, it will commit all its psychic energies to hate and envy, resentment and thirst for revenge ... Perpetual losers can be found on all continents. Among them the feeling of humiliation and the desire for collective suicide is increasing by the year. A nuclear arsenal is already available on the Indian sub-continent and in the Soviet Union. Where Hitler and Saddam failed – in achieving 'final victory', the final solution – the one who inherits their mantle may succeed.

This emphasises the point that starving and displaced people are not only an offence to human dignity such as to attract compassion and aid. Those more fortunate giving attention to the needs of the less fortunate is a matter of enlightened self-interest. Further, if we do not, we are likely to reap a harvest of violence and destruction for which, as I shall argue below, those of us living in First World comfort and security bear a heavy responsibility.

However, action stemming from enlightened self-interest can only emerge with a sophisticated degree of conscious awareness. There are profoundly important psychological driving forces which underpin human acquisitive behaviour which have their origin in our basic animal biology.

Biological imperatives and resource use

Much basic biological behaviour, like competition for the resources necessary for survival, enrichment and providing for a family, has become so culturally ritualized as to be almost unrecognizable. Such ritual, like behaviour in committees, or the way the hierarchical power structures of institutions are arranged, serves to make basic biological behaviour socially acceptable. Further, the acquisition of material resources is given the utmost prominence in our society. Arguably, the pressure to acquire more and more has never been greater; marketing to achieve this never more powerful. This forms an unholy alliance with basic – both male and female – unconscious drives, yet has little or nothing to do with the fulfilment of basic needs beyond a certain, relatively low level. Basic biological and psychological drives are being constantly activated at an unconscious level.

So it is, at some level, found to be acceptable – or at least inevitable – that we have some members of society in the North (First World countries like USA, UK, Japan) living in extraordinary opulence a few miles from others who retire to a cardboard box for a cold night, to say nothing of the starving billion elsewhere in the world. The operation of 'market forces' and 'freedom of opportunity' are the euphemisms given to an otherwise undignified scramble, with winners and losers, globally and locally, as there have always been throughout history – and indeed in the animal world.

Altruism – helping another at a cost to oneself – is not a normal part of animal behaviour. It can always be shown that there is self-interest somewhere underlying even the most apparently altruistic human behaviour (Dawkins, 1989). One altruistic mask that might be identifiable in the North–South (First World–Third World, developed–developing countries) relationship is the role of the voluntary aid agencies. More realistically, their roles might be seen as the vanguard of enlightened self-interest. Otherwise, the relationship is largely exploitative of the South by the North. This is rooted in history and past colonialism, and helps maintain the net resource flow (1989) of $51 billion from poor countries to the already rich (UNDP, 1991):

> During their colonial past the Third World countries experienced an improvement in living conditions that was sufficient to reduce death rates and thus initiated population growth. But the rise of living standards did not continue, because the wealth generated in the colonies was diverted to the developed countries, where it helped *their* populations become balanced. This process continues today, as many Third World countries remain colonized in the economic sense. This exploitation continues to increase the affluence of the colonizers and prevents Third World populations from reaching the standard of living conducive to a reduction of their rate of (population) growth (Capra, 1982).

Distribution of resources and economic growth

It is the generally accepted view that 20% of the world's population consumes 80% of the resources used (Meyer and Sharan, 1992). Yet, there is no sign of this imbalance being redressed. The more that perpetual economic growth is posed as a solution, the more it triggers the problem because of the dependence of countries of the North on the natural resources of the South, together with the continuing demand in the North for an ever higher standard of living.

There may be awareness amongst international politicians that, in the long term, perpetual economic growth is impossible but, despite this, the Brundtland Commission (Redclift, 1987), in producing the fudge of *sustainable economic growth*, tacitly recognized the unacceptability to First World electorates of foregoing the prospect of a continuing rise in material consumption. This short-term political realism has been perpetuated in the legally binding commitment in the UNCED Climate Convention to promote strong and sustainable economic growth.

Human population growth rates in the South are frequently considered to be at the root of current and future global environmental and resource problems. This does not bear serious examination as Capra (above) indicates, even though continued high population growth rates make problems of environmental degradation much worse. High population growth rates are a positive function of poverty. Most countries of the South are caught in a relentless strait-jacket of financial indebtedness to ones of the North – so much so that the average African country has an external debt greater than 100% of its GNP. This in turn inevitably results in the unsustainable use of renewable resources as pressure on land increases for the production of food, both for direct consumption and for cash. As the Brundtland Commission (Redclift, 1987) observed, poverty leads to environmental degradation which in turn leads to greater poverty – a vicious spiral. Even the international business community is beginning to recognize where responsibility for this chain of events lies:

> Environmental arguments for slowing human reproduction rates in the South pale before the truth that most environmental offences with global impacts are committed by rich minorities with hazardous production patterns, energy use and consumption styles (Schmidheiny, 1992).

and:

> Human resource development, which is essential for sustainable development, itself requires effective population management through programmes that recognise the linkages between poverty and population growth. (World Commission on Environment and Development, 1992; in Meyer and Sharan, 1992).

This First World responsibility can be exemplified in relation to greenhouse gases by examining regional outputs of industrial CO_2 emissions, together with outputs per head of population for two contrasting countries (Fig. 1a, b). The economic development of the North has been partly at the expense of unconsidered environmental costs. It is disingenuous to imagine that countries who have not been the beneficiaries of this growth will readily accept curtailment of their own prospects of development by a similar route without substantial support. It is even more unlikely that they will be prepared to help pay for the restoration of environments damaged in support of the economic growth of the North.

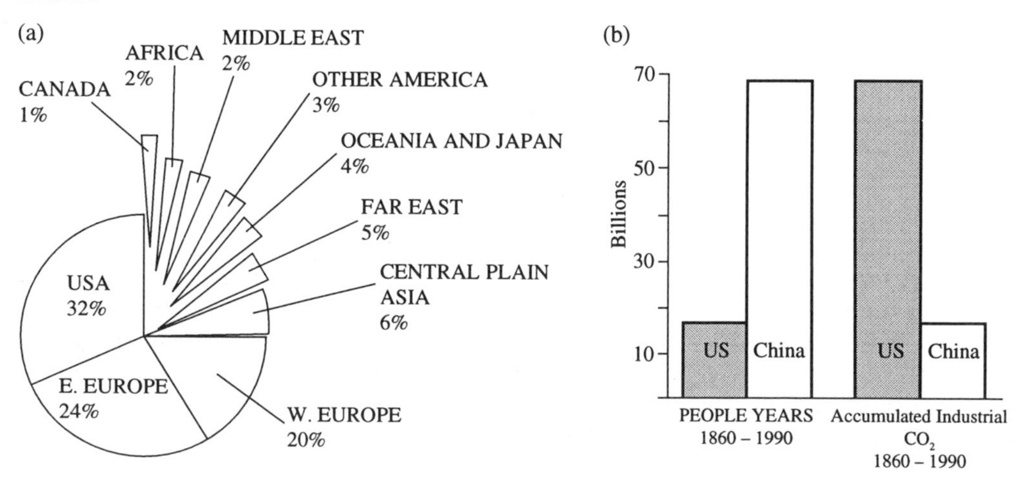

Figure 1. (a) Regional industrial CO_2 emissions 1860–1990 expressed as percentages of global total output (after Global Commons Institute, 1993); (b) accumulated population expressed in people years (number of people × average lifespan) and industrial CO_2 output in tonnes for USA and China, two contrasting large countries of North and South (after Meyer and Sharan, 1992).

Further economic growth involving, on the one hand, continuing local and global environmental costs and, on the other, disproportionate benefits to First World lifestyles is set to continue. The future consequences are unpalatable. The longer the present and prevalent course is pursued, the more difficult it will be to change it and the more costly will be the environmental repair.

The future

I have outlined why the sum of global human misery represented by deprived and hungry people is like a time bomb for those who are secure and well fed. It seems likely that there are two very different alternative responses from individuals and nations of the First World who have much to lose:

(i) Use force and physical limits to prevent Third World countries from emulating the First World.
(ii) Change direction in the First World to a sustainable economy and sustainable use of natural resources and assist Third World countries in copying it.

> Because I find the first alternative both immoral and unrealistic I am coming to the conclusion that sustainability is primarily a task for the North. (Weizsacker, in Meyer and Sharan, 1992)

However, there are two further problems which have to be faced:

(i) Accountability for what has happened.
(ii) Assumptions built into the most recent economic models

The starting point is the equal right of people to live and meet their basic needs for food, safety and security – to live in an environment within which there is the reasonable prospect of meeting higher order personal needs. This is *equity*. It follows that the only sound ethical starting point is that the value given to a human life is the same throughout the world. To what extent such a position is acceptable to a majority of Northern electorates is unknown.

However, it is apparent that at least some prominent economists believe that equal *per capita* rights are fundamentally unacceptable to Northern electorates (Meyer, 1994). Furthermore, an environmental economist (Fankhauser, 1992) has recently suggested in the context of estimating the cost of global warming damage that *statistical* lives in developed countries can be valued at $1.5 million while those in low income countries can be valued arbitrarily at $100 000. Although the author disclaims the idea that this means that the life of, say, a Chinese is worth less than that of a European, such assumptions, if built into global economic models, inevitably mean that disasters involving loss of life in the South are considered to be much (perhaps by an order of magnitude) less significant than those in the North. Similarly, 'impeccable economic logic' led a senior economist at the World Bank to suggest, prior to the UNCED Conference (1992), that countries in the South should commercially exploit their 'under-pollutedness' by selling pollution space to the overpolluted countries of the North (Meyer and Sharan, 1992).

Building such values into economic models under the guise of intellectual logic is a dangerous game. If such views are prevalent, they signal an unwillingness to make the radical change of direction in international economic policy that adherence to an acceptable ethical stance demands. The meaning of such a message will be clear to the South.

At the national level, the depth and extent of the response of governments to the commitments made under the Climate and Biodiversity Treaties remains to be seen. However, if the UK's initial response to the newly established UN Commission on Sustainable Development can be taken as representative, there is no sign yet of any radical changes of attitude or of direction. The UK Report to the Commission (Department of the Environment, 1994) is a bland and self-satisfied statement which suggests *inter alia*: 'Sustainable development does not mean having less economic development' (paragraph 12). The traditional economists' assumptions are clearly still firmly in place. Britain has no long-term strategy for dealing with the range of resource- and land-use issues that are involved in sustainable development (Baines, 1994). There also seems to be no understanding of the way in which development in the South is inextricably linked with UK sustainable development: 'The Government recognises the importance of encouraging sustainable economic growth in developing countries ... The UK's aid programme plays an important role ...' (paragraph 90).

The cost of the UK's aid programme is a tiny fraction of what might reasonably be required from this country if the commitments under Agenda 21 are to be met. This may reflect a perceived unwillingness in the electorate to support such a transfer of wealth, in line with statements made by economists. Although four prominent UK NGOs (Royal Society for the Protection of Birds, Friends of the Earth, Worldwide Fund for Nature UK, and Council for the Protection of Rural England) are represented on the Department of the Environment's 'UK Sustainability Round Table' to help inject political energy into the broad sustainability agenda, the RSPB Chief Executive has said:

> I think it would be really great if we could get a common set of principles across the development and environment organisations ... the issues of social justice that are coming up as a result of that I just cannot sell to my members. (Rawcliffe, 1994)

If members of conservation organisations cannot be sold these issues who can be?

In the final analysis, individual attitudes and decisions will be what count. How these turn out will depend upon how each of us resolve the internal conflicts between safeguarding our immediate short-term interests, our lifestyle and our material acquisitions, and our view of the future world for our children. The balance between personal decision-making and grassroots activism on the one hand and government leadership on the other is a subtle one. Maybe global environmental crisis is too far ahead in the minds of the older generation of decision-makers for them to promote change – an attitude of 'the world will see me out'.

Such an attitude, however understandable, is a counsel of despair. All of us can become the victims of despair. At such times, William Blake's 'doctrine of minute particulars' (Moore, 1987) is worth recalling:

> He who would do good to another must do it in Minute Particulars. General Good is the plea of the scoundrel, hypocrite and flatterer; For Art and Science cannot exist, but in minutely organised Particulars.

In addition to stating an ethical position, this can be taken in essence to mean that major changes are – and need to be – made up from individual small actions at an incremental rate. Thus tiny fragments of our local environment are worth our attention; what we do with our waste is significant; how we influence the local council makes a difference. Of wider significance is the need to address social injustice just as much at the local scale as the global, in the North as much as in the South, recognising that increased social injustice inevitably leads to further environmental degradation. Equity begins at home. Many of the achievements of conservation over the last two decades have resulted from small-scale, local action. Every now and then they are supported by significant shifts in national and international decision-making. Whether such incremental small actions and larger shifts will be sufficient to promote an acceptable quality of life for people throughout the world in 25 or 50 years time remains to be seen. The prospects do not, however, look good.

Conclusions

The following conclusions stem from the argument outlined above:

(1) Figure 2 summarizes the way in which the environment, with its characteristics we desire and upon which we depend, becomes squeezed between the expanding and conflicting pressures of acquisition of material wealth in the North and the results of deprivation in the South. Personal psychological attitudes, particularly defence and desperation, determine individual behaviour and are collectively played out through the political process.

(2) Our long-term civilized survival depends on moving fast towards equity. The South (Third World) knows about equity and collective international movement in this direction is going to be increasingly demanded. As Norman Myers noted in a speech to Biodiversity Challenge in 1993, 'Leaders in developing countries will not be persuaded by fine-grained scientific analysis, but by the example we show in our own back yard'.

(3) The notion that somehow local and global biodiversity can be maintained through minor adjustment to the *status quo* is fallacious. This does not mean that conservation achievements in terms of the establishment of national parks and nature reserves are unimportant. They are a vital part of the process of moving towards sustainable management of the biosphere. However, their survival depends upon sustainable management of the whole biosphere of which they form part.

(4) Equity, together with sustainable management of the biosphere, has radical and severe implications for prevailing lifestyles in the North. The increasing recognition of this is so far not reflected in political action towards a change of direction. Decision-makers appear to believe that equitable use of the world's resources is unacceptable to Northern electorates because the price in terms of lifestyle change is too great.

(5) The political entrenchment of growth economics, perhaps increasingly driven by the needs of multinational corporations, makes a radical change of international direction extremely problematic. The inertia in the present prevailing economic system is so great that it is difficult to imagine how moves towards an internationally sustainable economy will come about.

(6) The current improbability of acceptance of equity as the guiding principle of international relations and environmental management suggests a dire future scenario in which conflict will predominate over co-operation within and between North and South. The use of force and physical limits to maintain the *status quo* follows naturally from an unwillingness to change the nature of international relationships.

(7) Acceptability world-wide of an ethic of social justice would not on its own ensure the safeguarding of biodiversity and ecosystem functioning. Conventional measures of goals and ideas of betterment have to reach beyond those based on the accumulation of personal possessions. They have to embrace in a new way traditional cultural and spiritual values respectful of non-human life and the places it requires to sustain itself.

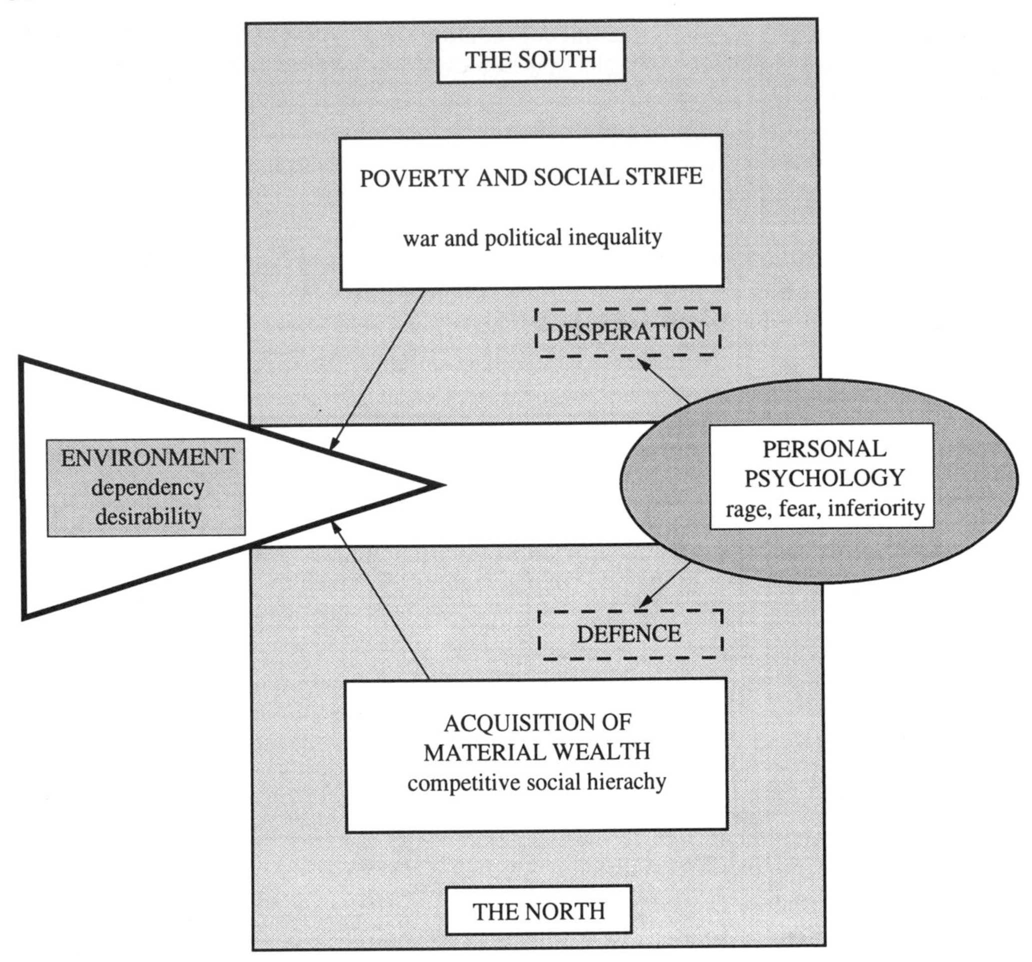

Figure 2. North–South political and environmental relationships. This represents a summary of the way in which the environment, with its characteristics we desire and upon which we depend, becomes squeezed between the expanding and conflicting pressures of acquisition of material wealth in the North and the results of deprivation in the South. Personal psychological attitudes, particularly defence and desperation, determine individual behaviour and are collectively played out through the political process.

(8) The major global ecological and environmental problems and how they can best be solved are generally well enough known. Where they are not, the precautionary principle should apply. Ecologists have both much to offer and still much to find out about how to protect and conserve species in their habitats, together with habitat re-creation and restoration. However, their role in tackling the larger, wider, more fundamental problems in focus here needs to be more vigorously political. They know quite enough to be so. Courage is often lacking.

(9) A lack of scientific knowledge is only an excuse, and not sufficient reason, for inaction in relation to environmental matters. Whether or not the problems of the global environment will be solved depends upon the balance of attitudes at every level – from the

individual to the international community. Therefore the empowerment of individuals and, in the North, a quantum shift of the spirit, are at the root of achieving biodiversity conservation.

Acknowledgements

I am indebted to Aubrey Meyer of the Global Commons Institute, 42 Windsor Road, London NW2 5DS, UK, for drawing my attention to key aspects of international economic policy, for providing me with source material, and for much stimulating discussion; and to Nigel Cooper for his helpful editorial criticisms.

References

Anon (1993) Brazil of the North. *BBC Wildlife* **11**, 11.

Baines, C. (1994) The Baines report. *BBC Wildlife* **12**, 62.

Capra, F. (1982) *The Turning Point: Science, Society and The Rising Culture*. London: Wildwood House.

Dawkins, R. (1989) *The Selfish Gene*. Oxford: Oxford University Press.

Department of the Environment (1994) *United Kingdom report to the commission on sustainable development*. London: DoE.

Editor of Nature (1972) The case against hysteria. *Nature* **235**, 63–5.

Enzensberger, H.M. (1991) The second coming of Adolf Hitler. *The Guardian* **9 Feb. 1991**, 23.

Fankhauser, S. (1992) Global warming damage costs: some monetary estimates. Working Paper GEC 92–29. University of East Anglia: Centre for Social and Economic Research of the Global Environment.

Global Commons Institute (1993) *Climate Change and The Precautionary Principle*. London: Global Commons Institute.

Goldsmith, E., Allen, R., Allaby, M., Davoll, J. and Lawrence, S. (1972) A blueprint for survival. *The Ecologist* **2**, 1–43.

Maslow, A.H. (1970) *Motivation and Personality, 2nd edn*. New York: Harper & Row.

Meadows, D.H., Meadows, D.L., Randers, J. and Behrens, W.W. (1972) *The Limits to Growth: A Report to The Club of Rome's Project on The Predicament of Mankind*. London: Earth Island Press.

Meyer, A. (1994) *IPCC Colonised by OECD Economists and 'Rights By Income'*. London: Global Commons Institute.

Meyer, A. and Sharan, A. (1992) *Equity and Survival: Climate Change, Population and the Paradox of Growth*. London: Global Commons Institute.

Moore, N.W. (1987) *The Bird of Time: The Science and Politics of Nature Conservation*. Cambridge: Cambridge University Press.

Odum, E.P. (1969) The strategy of ecosystem development. *Science* **164**, 262–70.

Pope, H. (1994) Death hangs over Black Sea. *The Independent on Sunday* **23 October 1994**, 13.

Rawcliffe, P. (1994) Stemming the tide: the changing nature of environmental pressure groups in the UK 1984–1994. Norwich: PhD Thesis, University of East Anglia.

Redclift, M. (1987) *Our Common Future: The Report of The World Commission on Environment and Development (the Brundtland Commission)*. Oxford: Oxford University Press.

Royal Society (1992) *Joint Communique on Population*. London: The Royal Society.

Royal Society for Nature Conservation (1989) *Losing Ground: Habitat Destruction in The UK: a Review in 1989*. Lincoln: RSNC.

Schmidheiny, S. (1992) *Changing Course: a Global Business Perspective on Development and The Environment*. Cambridge, MA: MIT Press.

Tickell, O. (1993) The green chainsaw. *BBC Wildlife* **11**, 11.
United Nations (1992) *Agenda 21: Rio Declaration*. Rio de Janeiro: United Nations.
United Nations Development Programme (1991) *Human Development Report 1991*. Oxford: Oxford University Press.

8

Christianity and human demographic change: towards a diagnostic ethic

SUSAN POWER BRATTON
Arts and Humanities, University of Texas at Dallas, Richardson, TX 75083, USA

Dialogue between Christianity and ecology over demographic issues is limited by four problems: a preference for the global and the apocalyptic, a preference for first order ethical approaches, lack of mutual comprehension of scientific and religious vocabularies and concepts, and inappropriate application of scriptural passages. Christian environmentalists and ethicists could enhance their understanding of demographic processes and respond more sensitively to demographic change by pursuing a diagnostic ethic that concentrates on regional data and identifies the appropriate scale of potential impacts; tests concepts on well-developed case histories; carefully identifies the specific issues, such as questions concerning women's rights and those concerning environmental degradation; accurately describes the concerns of the people actually involved; and recognizes that religion can either significantly hinder or greatly help appropriate social response.

Keywords: human demography; environmental ethics; Christianity; global issues

Introduction

The lack of dialogue between ecology and theology concerning the ethics and consequences of human population growth and demographic change is an example of the seemingly unending conflict between religion and science. This is unfortunate, because poor communication increases the possibility of undesirable environmental impacts and human misery accompanying ever-increasing human numbers. Different fields of study have different spheres of influence: ecology provides an understanding of the degradation of regional landscapes, and of the impact of human populations on natural resources, and provides information concerning the speed at which agricultural and industrial development can safely and realistically proceed; the social sciences, including demography, give us insights into how and why human populations change; theology and religion often mediate cultural responses and influence the way individuals and communities respond to such pertinent factors as marriage, child bearing, women's roles, contraceptive practices, respect for the land and community organization.

We usually either hold religion responsible for too much, or else overlook it entirely. Through recent European history, for example, Christianity has helped to maintain monogamous forms of marriage, but economic variables, such as the price of grain, or the availability of land for farming, have probably had more influence over birth and death rates, and the age of marriage. All these factors influence the rates of regional population growth or decline, and the effectiveness of human response to changing natural and social environments. Academic Christian ethics, however, have been slow to confront population issues. Although books such as my own (Bratton, 1992) *Six billion and more:*

human population and christian ethics, and collections of papers, such as the section on 'Population and Religion' in Laurie Mazur's (1994) *Beyond the numbers: a reader on population, consumption and the environment* are beginning to close the gap between the environmental literature and Christian responses to population questions, the way Christian ethics selects its subjects and analyses the issues may, in itself, have slowed the process of confronting population as a subject for study and discussion.

The purpose of this essay is to encourage further and more objective dialogue between ecology and Christian theology, particularly that of North American Protestants, in the area of human demography. Such dialogue can help us all to evaluate the potential impacts of population change for our era and for present and future generations, and can encourage the world's religious communities to assist in coping with change. Although many environmentally-oriented Christians and Christian organizations, such as the World Council of Churches, are already grappling with population questions, the theological discussion remains dispersed and often does not fully engage the ecological information available. This essay will outline four problems in developing a sophisticated Christian approach to human demographic change: a preference for the global and apocalyptic, a preference for first order approaches, lack of mutual comprehension of scientific and religious vocabularies and concepts, and inappropriate application of scriptural passages. The conclusion of this paper will propose that rather than seek very general, first order ethical models, Christianity will be able to deal better with the realities of human reproduction if it pursues a diagnostic form of ethics.

World demise or regional responses

A major difficulty in developing ethical Christian responses to demographic change is the tendency for both Christians and environmentalists to define human population problems as primarily global. As Boyer (1992) has pointed out, some environmentalists and many Christian denominations have an apocalyptic view of human history. On the Christian side, premillenialism, an eschatological position (theological view of the future) especially widespread in the US, expects a worldwide tribulation, including massive disruption of the earth's environment. God will then renovate the planet and the evil will be undone. Other Christian projections of the *eschaton* or end of time, may be less dramatic, but still anticipate some form of planetary change. On the environmentalist side, there is concern for a major failure of the earth's life support systems, initiated by a combination of human environmental abuses, including damage to the ozone layer, increasing carbon dioxide in the atmosphere, clearing of tropical forests, and overuse of agricultural lands. Most environmentalists, of course, do not interpret these as resulting in divine judgment nor do they expect divine intervention to heal them. Apocalypticism encourages Christians to do nothing, since the problems are viewed as symptoms of a pre-planned global cataclysm. Environmentalists, in contrast, tend to over-react, and to become so concerned about population increases that the reproductive rights of women and families are subordinated to the need to 'save' the planet.

The notion of a 'global population bomb' implies human demographic processes are universal phenomena. This is not the case. Different regions have different population growth rates, and are experiencing different types of population change. Although there can be little doubt population increases may stress regional resource availability and lead to damaging environmental management practices, the local social and environmental context is important in determining whether an increase or decline in human population or a change in age structure is detrimental. Further, the 'global bomb' concept concentrates on total numbers. Ethically important questions, such as the impact of large family size on maternal health or relative use of resources may, therefore be overlooked. The vast majority of ethical issues concerning human demographics cannot be addressed on a worldwide basis. Not only is the regional context critical, but the cultural, economic and gender contexts are as well.

Since Christianity considers itself a religion open to all peoples of the earth, Christians tend to err on the global side, when considering questions of scale and context in social ethics. The desire to seek a universally right Christian path can blur the differences between peoples living in widely varying social settings, or physical environments. In the case of human demography, Christian ethicists need to differentiate between regions, recognize the way demographic change influences people of different social classes, discern the differential impact of some changes on people of different ages and genders, and understand that family structure varies with cultural history and economic pressures. In addition, Christian ethics should carefully distinguish immediate human and environmental needs from projected cosmic outcomes.

First, second and lower order ethics

Shrader-Frechette and McCoy (1993) recently defended the entire science of ecology, which has always had difficulty developing predictive theories broadly applicable to environmental problems. The great diversity of living organisms within the earth's ecosystems, and the variable physical conditions which prevail on the planet inhibit the development of 'exceptionless empirical laws and a deterministic general theory' (p. 279). They argue, however, that ecology provides useable natural history data and rough generalizations that may guide ecological decision making. This means that ecology may be most useful in actual problem solving when it takes a 'bottom up approach' and thoroughly investigates specific cases in detail.

Christian ethics, in contrast, is based primarily on sweeping, first order principles, such as 'you should love your neighbour as yourself'. Unlike many tribal or folk religions, Christianity is not tied to a specific region of the earth and much of its theology has developed independently of landscapes or local ecosystems. On one hand, this gives Christianity great cultural flexibility – a religious movement initiated by rural Galilean peasants in the first century C.E., lives on, not just in contemporary Rome and Jerusalem, but also in Los Angeles, London and Seoul. Christianity, however, has not produced an accumulation of laws or taboos concerning human residence in specific environments. Although there are Biblical passages and mandates encouraging care of the earth, a majority of these describe specific environmental features and problems in the Middle East. Two thousand years of western philosophical influence have kept academic Christian theology very abstract, and have fostered the search for the religious equivalent of empirical models and general theory. As Christianity has fragmented into diverse denominations, civil law has displaced church law. The once extensive land holdings of Christian communities, such as monasteries, are now much reduced. Christian ethics has thus lost much of its contact with actual environmental legislation and much of its direct authority over environmental questions.

The conclusions of Schrader-Frechete and McCoy (1993) suggest that one of the difficulties in developing a Christian demographic ethic is that Christianity, with its preference for first-order ethics, is attempting to draw on fields such as ecology, which may be most ethically productive when investigating specific cases. I have suggested (Bratton, 1993) that the Christian eco-feminist literature is often difficult to apply to actual environmental issues, because it favours abstract theological models, such as those utilized by process theology, or deals primarily with questions, such as those concerning gender, social, and biological hierarchies, only in general terms. Since understanding the sources and impacts of human population processes must rely on diverse information from ecology, economics and sociology, a sound Christian ethic can only develop after extensive consideration of individual cases. In order to pursue dialogue with ecology, Christianity must take such first order principles, as 'thou shalt not kill', or Biblical instructions to 'till and preserve' the garden, and interpret them in second and third order settings.

Closing the gap between first and lower order ethics is a common difficulty for Christians who are also environmentalists. They must link the more general Christian texts and concepts, to the specific cases, which are often described in scientific terms. A simple example is the question of whether a group of poor farmers, with a growing population, would be doing the wrong thing if they cleared a patch of forest harbouring one of the last populations of an endangered bird. A statement encouraging Christian 'stewardship' of the environment does not resolve the conflict. The Bible does not discuss the international red book or the genetics of declining species. The Bible does, however, credit God with establishing the diversity in creation. Further, the Book of Job and the Psalms indicate that God gave each species a home (or habitat). There are also numerous passages about aiding the poor. A thorough Christian analysis of the issue would attempt to determine community responsibility for the critical habitat, and the source of the social and political circumstances forcing the clearing of all available land. The final resolution of the issue involves more than one ethical principle and a detailed examination of the available economic options for the community.

Vocabulary and concepts of resource availability

A third difficulty for Christian ethics is that religious thought often addresses questions of interest to other fields, such as ecology, but with a different vocabulary and different priorities. One of the major concerns raised by ecologists is the concept that the earth's natural resources are finite and therefore there may be a limit to the number of humans the earth can support. Modern ecology has stressed the idea of carrying capacity and has proposed that successful species – not just human communities, but deer, sharks, and frogs – do not long exceed their available food supplies, and other necessities of life, such as space and water. The thought of ecologists is, in fact, quite deterministic: if a population frequently or for an extended period exceeds its resources, it will deplete those resources, which will result in a population collapse, and perhaps long-term damage to the resources themselves.

This type of model has been frequently employed for the earth's human population, but due to the complexities of resource distribution within and between human communities, most ecologists are wary of predicting some specific total is more than the planet can sustain. Both social and natural scientists express a continuing concern that human population growth rates are too great and this will lead to inevitable disaster, at least in terms of regional famines, deforestation and increasing difficulties with pollution.

Although the vocabulary is different, Christian theology also deals with questions of resource availability and of the potential for disaster. The most central concept is that of providence: God's continuing care and provision, not just for humans, but for all life on earth, and the cosmos as a whole. Confusing not just for scientists, but also for Christians themselves, is the tangle of popular and formal theological approaches to providence. The general public is more likely to be aware of culturally widespread models such as the 'health and wealth gospel' of some television evangelists, that it is to be influenced by academic ruminations about whether God is responsible for the evil in the world, or whether God takes risks. (For a formal theological discussion of providence, see, e.g., Helm 1994.) Popular religion in North America is often innately optimistic, ensuring the person in the pew that 'God will provide'. Unfortunately politicians and voters hear and respond to such popular views, and this, in turn, affects how human cultures respond to environmental and demographic change.

Let me suggest that the concept of ecological limits does not in any way undermine the Christian concept of providence. However, Christians have not been careful enough in distinguishing whether certain types of divine provision are infinite or finite. Divine love is infinite and so is divine mercy. The agricultural productivity of a region, or the water reserves in an aquifer are not. Some Christian environmental commentators have adopted the thought of economists such as Julian Simon (1977, 1981) and have suggested that human innovation is infinite (or close to it) and will eventually resolve any environmental problems created by human population growth. Beisner (1990), for example, considers life, liberty and property to be 'paramount' principles, that are 'Biblically sound', and proposes that a free market economy 'within the limits of God's moral law' can solve most of our environmental woes.

In contrast, Betsworth (1990) has proposed that the popularity of 'the gospel of success' in America is based partially on self-deception. He notes that some of its proponents, such as Andrew Carnegie, did not recognize the gifts they had been given by others (or by God), and concludes: 'the real story has to do with being providentially placed in a world with fewer limits than most, and with being given gifts that are wholly undeserved...' Responding to environmental need, and executing environmental correction often require the input of capital or labour, as well as the will to deal with the problems. 'Hope' cannot be programmed into a computer. Christians must recognize that the scientists deal with what is immediately known about natural environments, and the mathematical models they develop are based on finite resources. Christian demographic ethics must combine a belief in divine good will, with the harsh realities of a world 'characterized by contingency, contradictions and limits' (Betsworth, 1990).

In engaging and articulating ecological concerns, Christian thought needs to further confront some of its own internal contradictions. What, e.g., does 'hope' mean in the poorer regions of the world? Is 'providence' equally bestowed on the human and the non-human? Do Biblical concepts of 'providence' refer primarily to the daily blessings of food, clothing and shelter? To what extent is providence extraordinary? Can humans interfere with and divert providence? To what extent are the times when there are not enough provision the result of divine will as opposed to human injustice? How much power does God actually exercise?

Ancient texts and 'more is better'

A last issue in developing ethical dialogue between Christianity and ecology is the dependence of Christian perspectives on human reproduction and resource availability on the interpretation of Biblical texts and other historic sources. Ecologists may suspect that ancient religious writings are not adequate to answer the complex questions produced by modern industro-technical culture, and that the Bible would be better relegated to courses on the world's great literature, rather than being employed to address ecological or economic crises. Both Christians and non-Christians, for example, often unfairly assume that the Bible is unrelentingly pro-natalist (favouring high birth rates). Ecologists may believe it promotes 'fertility cults for cowherds' and is utterly irrelevant when discussing human reproduction in contemporary mega-cities.

Although careless interpretation, which ignores the original cultural settings and the central theological themes of the literature in the Christian canon, may give the impression that Biblically-based Christianity is impervious to the difficulties presented by human population growth, a socially and theologically sensitive reading of relevant texts can actually present a very different picture. An example is the instruction in Genesis 1:28 to humans to 'be fruitful and multiply and fill the earth'. Interpreted literally, and without regard for the passages around it, this reads like an ethical imperative to produce as many children as possible, and to use every last hectare of arable land on the planet. If we recognize, however, that the divine instructions to be fruitful and fill the earth were first given to the birds and the sea creatures (including the oceanic monsters), we will also recognize that treating the Genesis texts simply as ethical imperatives, implies that, say, a massive sea lamprey invasion of Chicago would fulfil divine wishes. Worse yet, if 'more is better', the instructions place humans in a reproductive race with rock doves, rats and beetles to fill the earth before another taxon does.

A better reading of this passage, following the analysis of Old Testament scholar Walter Brueggemann (1982), is to interpret the instructions to 'be fruitful' first as a blessing, and as a blessing shared with non-human creation. The passages suggest that the animals respond immediately 'not because of coercion but because creation delights to do the will of the creator'. Bratton (1992) suggests:

> While the blessing does impart generative power and fertility, its intent is reproduction in balance, springing joyfully forth to produce the well-being God continues to weave into the entire created universe. The blessing is part of what makes creation 'very good' and very beautiful in the eyes of God. Human population growth has no mandate to damage or desecrate the cosmos, nor is it God's intention, although the act of childbirth itself is painful (Gen. 3:16), that human reproduction be a source of societal sorrow and suffering.

In addition, since the Bible was written in a time when people lived in extended families in rural settings, without modern medicine or modern economies, we must also be careful to contextualize passages about child bearing and family structure properly.

Towards a diagnostic ethic

A potential solution to the miscommunications and inappropriate analyses described above is to develop a 'diagnostic' approach to Christian demographic ethics. Rather than attempting sweeping 'first order' or global solutions, demographic questions should be subjected to a diagnostic process to determine the appropriate scale, sub-issues and information bases. Due to the technical, and in some cases legal nature of case materials, ethical discussion should incorporate not just theologians, but also ecologists, sociologists, physicians, economists, policy analysts and others who can help in clarifying the sources and impacts of demographic change. Further, ethicists do not have some special 'god-given' authority to make decisions concerning life and death, nor, in many cases, do they necessarily have direct experience with the human or environmental damages involved. The current trend in medical ethics, whereby philosophically trained ethical 'experts' are replacing care-givers and medical researchers in teaching and professional publication should be avoided, and a team approach pursued so that diverse perspectives, professional expertise and *sensitivities* can be incorporated in demographic problem solving.

The first question is one of scale: At which level should an issue be properly addressed? How is it best defined both in space and in time? Christians who see God's role in history as central may prefer to generalize. This however ignores the importance of the regional or local context. Many African and South American nations now have very young populations, for example, while some European nations have relatively old ones. The potential negative impact of large family size on both child survivorship and maternal health is primarily a problem of the poor. A discussion of world population trends must consider both Germany and Zaire, both Canada and Brazil.

Second, the ethical responses should be based on or tested through well-developed case histories. Most dependable ecological and economic analysis is likely to be second or third order. Although theoretically oriented ethicists might prefer to discuss 'all women' or 'all families', it is more accurate to investigate 'African farming women in semi-arid lands', or 'women in southeast Asian countries with expanding economies'. Differences in environmental circumstances may make major differences in outcomes. General Christian precepts such as the sacred nature of human life or God's love for all people, must be carefully translated into lower order cases that may not be covered either by detailed Biblical discussion or by historic Christian theological commentary. Christians often discuss subjects like monogamous marriage and fidelity in marriage in terms that apply to all Christians. The advantages and disadvantages of early or late marriage, or large or small families are to some degree specific to the social context, making it impossible to devise a 'right' marital and reproductive strategy for all times and circumstances. Although divine love for humankind is eternal, family sizes and structures should and will vary.

Third, what are the specific issues, and what types of information are relevant? Among Christian ecofeminists, population has received surprisingly little commentary in the academic literature. However, women's or reproductive rights are almost always a component of population questions concerning population growth or birth rates. Human demographic change causes at least minor economic and environmental impacts in all types of societies from hunter-gatherer to agricultural to industrial. What will the impact be on availability of agricultural land if the population doubles? Given the best demographic data and economic projections may not be accurate, what are the most likely deviations? In some cases of demographic change environmental factors may be far more important than in others, so environmental data should be differentially incorporated in ethical discussion.

Fourth, religious communities should recognize that religion can either significantly hinder or greatly help appropriate social response. Christians must reflect on the human and environmental impacts of their teachings, doctrines, activities, and social services.

Christians also should repeatedly ask if their responses truly relate to actual social realities. The first step is to question what the impact of Christian teachings has been in the past, and how it is influencing current marriage and child bearing patterns. Christians should determine if child survivorship within their own congregations is good or poor, and if the health needs of women, including new mothers are being met. Is Christian response really helping children? Does Christian response have an impact on patterns of teenage pregnancy, the number of children completing secondary school or on the rate of infection of sexually transmitted diseases?

Lastly, diagnostic ethics should ask the families (and specifically the women) what their concerns are. Christians should at least consider the perceptions of their own church members, and of their own communities. Further, if a solution to a specific problem is proposed, Christian demographic ethics should be informed enough to know whether the people actually involved are willing to participate without coercion.

The ultimate goal of 'diagnostic ethics' is to approach demographic change in a way that is faithful to Christian 'first order' principles while thoroughly engaging ecological and economic reality. Diagnostic ethics should 'hear' the voices of the suffering, while 'seeing' environmental degradation and understanding its roots. Diagnostic ethics should then be able to 'speak' to the problems and conflicts in ways that all the 'stake holders' will understand.

References

Beisner, E.C. (1990) *Prospects for Growth: A Biblical View of Population, Resources, and The Future.* Westchester, IL: Crossway Books.

Betsworth, R.G. (1990) *Social Ethics: An Examination of American Moral Traditions.* Louisville, KY: Westminster/John Knox Press.

Boyer, P. (1992) *When Time Shall Be no More: Prophecy Belief in Modern American Culture.* Cambridge, MA: Harvard University Press.

Bratton, S.P. (1992) *Six Billion and More: Human Population Regulation and Christian Ethics.* Louisville, KY: Westminster/John Knox Press.

Bratton, S.P. (1994) Ecofeminism and the problem of divine immanence/transcendence in Christian environmental ethics. *Science and Christian Belief* **6**, 21–40.

Brueggemann, W. (1982) *Genesis.* Atlanta, GA: John Knox Press.

Helm, P. (1994) *The Providence of God.* Downers Grove, IL: InterVarsity Press.

Mazur, L.A. (1994) *Beyond the Numbers: A Reader on Population, Consumption and The Environment.* Washington, DC: Island Press.

Shrader-Frechette, K.D. and McCoy, E.D. (1993) *Method in Ecology: Strategies for Conservation.* Cambridge: Cambridge University Press.

Simon, J. (1977) *The Economics of Population Growth.* Princeton, NJ: Princeton University Press.

Simon, J. (1981) *The Ultimate Resource.* Princeton, NJ: Princeton University Press.

9

Forging a biodiversity ethic in a multicultural context

DAVID R. GIVEN
David Given & Associates, 101 Jeffreys Road, Christchurch, New Zealand

New Zealand belongs to the Pacific region, a part of the world where human impacts have been both very recent and extreme in their effect. The New Zealand natural environment is rich in endemic taxa, but these are poorly equipped to cope with the effects of invasion by humans and exotic animals and plants. Polynesian immigrants brought to New Zealand a distinctive world view which gave rise to both tribal traditions and living traditions of the Maori. The resultant environmental ethic emphasises guardianship and stewardship, establishment of the right to use a resource, kinship obligations, and a balance between pairs of opposites. Nineteenth-century European colonists were ambivalent in their view of the environment, although a world view which emphasises 'dominion' has tended to dominate. Two recent developments which are important factors in development of a multicultural biodiversity ethic are the enactment of the Resource Management Act 1991 and legal recognition of the principles of the Treaty of Waitangi. The intersection of these developments provides an opportunity to develop a new approach to environmental ethics especially in conceptualising 'significance', consultative processes, and developing a holistic and ecocentric use of resources.

Keywords: biodiversity ethic; Maori; New Zealand; Resource Management Act; tribal tradition; indigenous biota; stewardship

The Pacific environment

The Pacific region, of which New Zealand is part, includes more than 1000 islands and archipelagos. Some are oceanic in origin, having never been part of larger land masses. Other islands such as New Caledonia and New Zealand are fragments of the former southern continent of Gondwana. The region is notable for the diversity of its biota and high levels of endemism both at a regional and local level (e.g., Dahl, 1986; Murray *et al.*, 1988; Johnson and Stattersfield, 1990; Pimm *et al.*, 1995). Over 200 islands are documented as having notable 'natural richness' at a scale of global significance (Dahl, 1986).

Outside of Antarctica, this is the last major region of the world to be colonised by humanity. The spread of people into the Pacific, started with the northern Solomon Islands about 30 000 years BP and Micronesia about 2000 years BP (Dodson, 1992). The recent arrival of people has had devastating effects on the biota of Pacific islands (e.g. Pregill 1988; Flannery and Calaby, 1990; Flenley *et al.*, 1991, Given, 1994a; Pimm *et al.*, 1995). This has partly been due to the destruction of indigenous biota through both harvesting and habitat loss (e.g. Dahl, 1986; Pregill, 1988; Flenley *et al.*, 1991) but also through introduction of exotic animals and plants, some of which have replaced elements of the indigenous biota of the Pacific (e.g. Atkinson, 1985; Murray *et al.*, 1988; Flannery and Calaby, 1990; Whistler, 1991).

The spread of humans across the Pacific has been likened by Flannery (1989) to a plague, with the conclusion that, 'ironically these islands are among the most devastated on Earth! Some are already damaged beyond repair and all are sliding inexorably towards oblivion'. This statement is supported by the analysis of avian extinction rates in the Pacific region by Pimm *et al.* (1995). They conclude that the first colonists exterminated about half the species on each island group. Extinctions continue to the present day and many surviving species are so critically endangered that we know neither whether they still survive or how to save them. Similar scenarios are emerging for other groups of biota such as reptiles (Pregill, 1988) and landsnails (Murray *et al.*, 1988).

The New Zealand natural environment

Within the Pacific region, the biological diversity of New Zealand is of particular global interest for three reasons. The first is that it has very high levels of endemism, for instance 82% for vascular plants (Given, 1980), 57% for land birds (Robertson, 1985), and up to 90% for some invertebrates (P. Johns personal communication). The second is that some of these endemic taxa represent clades which have retained primitive characters (colloquially known as 'living fossils'). Examples include the fern *Loxoma cunninghamii*, the tuatara (genus *Sphenodon*), and the bird families Apterygidae (kiwi), Callaeidae (New Zealand wattlebirds), and Xenicidae (New Zealand wrens). These taxa evolved and survived in the absence of browsing and grazing mammals. The third is that Polynesian voyagers probably reached New Zealand only about 1000 years BP, making this the last sizable land mass of the temperate or tropic regions (area 240 000 km^2) to feel the impact of humans (Stevens *et al.* 1988; Patterson, 1994).

When people and the animals and plants which accompanied them arrived, indigenous biota proved to be extraordinarily poorly equipped to cope with the effects of invasion by this influx of exotics (Wardle, 1991). Polynesian rat introduced by Maori voyagers, later rat introductions, deliberate establishment of deer, goats, pigs, and Australian opossum, and spread of many hundreds of invasive weeds have each had a negative impact on the indigenous biota of New Zealand (Atkinson, 1985; King, 1990). Three indicators reflect the magnitude of human impacts on New Zealand biodiversity. It has been estimated that 70% of New Zealand was forested when the first Polynesian people arrived, dropping to 55% by the mid-19th century when widespread European colonisation commenced, and that currently about 27% is in forest (Given, 1980; Wardle, 1991). The number of exotic weeds established in New Zealand approaches the total number of indigenous vascular plants (Webb *et al.*, 1989). Australian opossums, originally introduced as the basis for a fur trade now number between 60 and 70 million individuals and are the primary cause of widespread dieback throughout New Zealand forests (King, 1990).

The Maori context

The Polynesian immigrants to New Zealand brought with them a theory of the universe in which they have a special relationship with the land and its biological diversity as *tangata whenua*. In a literal sense, this implies that Maori are 'people of and from the land', and it embodies the concept of interdependence between people and environment, creating a sense of belonging to nature.

One thousand years of settlement by Maori, saw the development of *matauranga*, which embodies the special skills needed for survival. It includes the concept of *kaitiakitanga* (guardianship and stewardship), exercised through *mana whenua* (customary authority exercised by a *whanau* or *hapu* social unit in an identified geographic area). Essential and highly prized resources constitute *taonga* or treasures of the world of nature. Vital to the concept of *kaitiakitanga* is the role played by *kaitiaki* (Patterson, 1994). These are the many spiritual assistants of the gods, including spirits of the deceased. They are spiritual minders of the elements of the natural world, including *taonga*. For Maori, the *mana* (power and authority derived from the gods) must be preserved for each element of nature as part of their *tapu* (sacredness). The environment is not simply a collection of resources to be exploited but is a community of related being, all of them linked to humanity by ties of kinship, all important and all needing protection through *kaitiaki,* and demanding respect through *tapu* and *karakia* (prayer). When *mana* is depleted, the *tangata whenua* in their role as *kaitiaki* must do all in their power to restore the *mauri* (vital life force) of *taonga* to its original state.

Kinship obligations (*whanaungatanga* and *manaakitanga*) are important in a Maori world view. These extend beyond just human kin to include obligations towards the environment; they are reciprocal with benefits extending also from the environment to humanity. Hospitality, based on a notion of reciprocity which in Maori is *utu* (the law of a corresponding and proper return), is fundamental not only to human relationships but to an ethic of nature (O'Connor, 1994). Hospitality is expressed as a two way love (*aroha*) in which others (human and non-human) have their established place in the universe. Furthermore, in Maori terms we are often held to be responsible for states of affairs that we have not brought about, and 'if we are to claim any *mana* over our environment and without *mana* we have no standing, no place, no access to "resources" – we must accept the associated, unconditional responsibilities of care and protection' (Patterson, 1994).

A particular subtribal or family social group with *mana* over a particular geographic area, natural feature or resource, are obliged to take their responsibilities seriously. Not to do so may result in failure of the resource through loss of life-sustaining capacity of land and sea. Failure to carry out *kaitiakitanga* may also result in the dissolution of the local *whanau* or *hapu*.

There are tensions within the Maori system. The traditional narrative of the separation of *Papa*, the Earth Mother and *Rangi*, the Sky Father, embodies elements of tension between the demands of kinship and the need for survival and growth (Patterson, 1994). This requires interacting, 'in a positive and *creative* way, making up for the harm we do by enhancing the environment in appropriate ways, as Tane did with his parents' (Patterson, 1994). The right to use resources is conditional; it depends on respect of *tapu* and the establishment of *mana*, always with the reminder that people have to respect the environment in which they live. As Patterson emphasises (p. 402) Maori, 'rather than seeing a respect for *tapu* as a matter of *following rules*, see it as *being a certain sort of person*'.

Traditional Maori technology is simple, but on the arrival of Europeans there was quick adaptation of European techniques and concepts to their needs. Nevertheless, Maori manage to maintain a dynamic balance between traditional and technological living and thinking, and current Maori ethics is in many important respects a traditional ethics (Patterson, 1994).

Five key aspects of Maori environmental values are noted by Patterson (1994). The first is that there is a distinction between *tribal* traditions which have their roots in the distant past but vary significantly from tribe to tribe, and *living* traditions which continue to evolve through time, including the period of European settlement. As a result, in a strict sense there is no one Maori world view, but a series of related world views and environmental values across related *iwi* or tribes (L. Morris personal communication). A second aspect is that Maori traditions are largely *oral*, and much has not and never may be committed to paper. Written versions have no special standing among Maori and much may never be written down because the traditions are seen as *tapu* (sacred or forbidden). A third point is that historical and ethical aspects of Maori narratives are connected through the ethical importance of *ancestral precedence* which provides models for ethical behaviour today. The fourth aspect is that although traditional *kaitiakitanga* over the environment may not be open to outsiders, it is also a virtue which is open on a voluntary basis to all who take it on themselves to care for the environment.

Lastly, emphasis must be given to *balance.* Although there is no 'important' word for balance, the concept is implicit in the Maori world view which is seen as consisting of polar pairs. The ideal is not to eliminate one of the pairs as undesirable or evil. Rather it is to seek the appropriate balance between the polar pairs. A result of this is the view that no one perspective is valid at the expense of any other, and to distinguish needs from mere wants (Patterson, 1994).

The European context

The earliest European settlers in New Zealand were sealers, whalers and timber millers whose activities were based on resource 'mining'. New Zealand was seen as an opportunity for uncontrolled exploitation of resources, well-illustrated by the catastrophic loss of marine mammal populations, principally New Zealand fur seals (*Arctocephalus forsteri),* within a short number of years (McNab, 1907; R. Taylor personal communication).

By 1840 the initial period of itinerant settlement had been succeeded by planned immigration from Europe. Of particular importance was the widespread establishment of Christian missionary settlements, and the orderly system of colonisation established by Edward Gibbon Wakefield (Hight and Straubel, 1957). The Wakefield system which resulted in regional colonisation based on Wellington, Nelson, Wanganui, Christchurch, Dunedin and Invercargill, was essentially a transplanting of church-based English society. It had profound effects on indigenous biota as species and their habitats were systematically replaced by the biota of Europe. Landscape transformation was deliberate, sudden and profound, especially in lowland regions, where it was essentially complete by the 1880s.

Many immigrants saw New Zealand as a place of new opportunities. Pastoral farmers, in particular, saw the existing forest, swamplands, scrub and tussock grasslands as an impediment to their plans for development (Hight and Straubel, 1957). Goldrushes, the development of coal mining and other extractive industries, and development of a comprehensive transport system led to an accelerating destruction of indigenous ecosystems in all but the inaccessible parts of New Zealand (Given, 1980; Wardle, 1991).

However, even during the initial stages of European settlement, concerns were expressed about the rate and degree of transformation of landscapes and ecosystems, and its effect on indigenous species. In 1856 the Austrian geologist, Ferdinand von Hochstetter, is reported to have commented unfavourably on the destruction of native forest close to the young settlement of Auckland. Thomas Potts (1882) noted with concern that, "timber clearing and burning progresses cheerfully on Banks Peninsula. We denude mountain and valley of forest growth, professing to believe that we are developing the resources of the country".

It is not surprising that the European view of indigenous biodiversity in the 19th century was ambivalent. The settlers had many varied motives for emigrating, and came from a great variety of backgrounds. Nevertheless, a common threat throughout the major period of initial European settlement was the strong influence of both evangelical and catholic Christianity. Anglican, Methodist and Presbyterian influences were strong and are reflected in much of the legislation on which the development of New Zealand and the use of its resources has been subsequently based (Laurenson, 1972; Glen, 1992).

New Zealand's major period of European colonisation and development has occurred at a time of emphasis on a doctrine of religious *dominion*, used as justification to alter, develop, change and destroy the pre-existing indigenous environments and dispossess indigenous people – all part of a world view in which resource use is primarily to promote the good of a few (Parkin, 1986). That ethic has guided much national development over the past 150 years but there are signs of groundswell of change, or at least recognition that thinking must change (Howell, 1986; Given, 1991).

The Resource Management Act

A major revision of New Zealand conservation and resource use legislation resulted in enactment of the Resource Management Act 1991. This consolidated much of the previous resource legislation and coincided with a major devolution of local government into a system of district and regional authorities. The Resource Management Act (hereafter referred to as RMA) places emphasis on environmental outputs rather than prescriptive control of resource use. The Act furthermore gives legislative recognition to a number of ethically-based concepts including recognition of intrinsic natural values as a matter of national importance. It also recognises the role of the community in shaping policy on resource use.

The primary purpose of the RMA, 'is to promote the sustainable management of natural and physical resources', sustainable development being defined as, 'managing the use, development and protection of natural and physical resources in a way, or at a rate, which enables people and communities to provide for their social, economic and cultural wellbeing and for their health and safety while –

(a) Sustaining the potential of natural and physical resources (excluding minerals) to meet the reasonably foreseeable needs of future generations; and
(b) Safeguarding the life-supporting capacity of air, water, soil and ecosystems; and
(c) Avoiding, remedying or mitigating any adverse effects of activities on the environment'.

The Act also sets out (sections 6–8) a hierarchy of matters which the Act shall, 'recognise and provide for' (section 6), 'have particular regard to' (section 7), and 'take into account' (section 8). Section 6 is particularly important. It states that particular matters are of national importance:

(a) 'The preservation of the natural character of the coastal environment (including the coastal marine area), wetlands, and lakes and rivers and their margins, and the protection of them from inappropriate subdivision, use and protection;

(b) The protection of outstanding natural features and landscapes from inappropriate subdivision, use and development;
(c) The protection of significant areas of indigenous vegetation and significant habitats of indigenous fauna;
(d) The maintenance and enhancement of public access to and along the coastal marine area, lakes, and rivers;
(e) The relationship of Maori and their culture and traditions with their ancestral lands, water, sites, waahi tapu, and other taonga.'

A strong ethical component is apparent, particularly in the use of value-laden terms such as *significance*, *enhancement*, and *outstanding*. This is echoed in section 7 which recognises, 'the maintenance and enhancement of amenity values', 'intrinsic value of ecosystems', 'heritage values', 'maintenance and enhancement of quality of the environment', and 'kaitiakitanga'.

The ethical aspect of RMA is reinforced by provisions (especially section 96) which recognise the role of the community and a wide range of interest groups in shaping environmental and resource use policy. Section 96 allows any member of the community at large to make a submission in support of, or objecting to, an application for a resource consent. This has potential to lead to a much more comprehensive consultative process than has existed under former legislation. In addition, general regional policy on resource use is guided by Regional Policy Statements in addition to formal regional and district planning documents, with requirements for comprehensive community consultation at the formulation stage.

The Treaty of Waitangi

The Treaty of Waitangi (Appendix A) was drawn up in 1840 as a result of British Government determination to secure Maori acceptance and cooperation in establishing New Zealand as a British colony. Over a period of 7 months more than 500 chiefs signed the Treaty. The Treaty itself consists of three clauses dealing with sovereignty, powers and ownership. An important clause, inserted by James Busby, guaranteed Maori possession of their lands, forests, fisheries and other prized possessions. The Treaty recognised that in exchange for settlement rights, Maori would have their natural rights as *tangata whenua* upheld; for Europeans the Treaty would enable peaceful emigration and settlement under the New Zealand flag.

Despite some initial concerns, both Maori and European settlers were generally in favour of the Treaty in the belief that it would ensure a number of necessary benefits:

- proper control of growth in trade;
- controls on how and where new immigrants settled;
- a common legal system for both Maori and European; and
- control on land sales.

Initial advantages conferred to both races diminished as the New Zealand administration passed to a Settler Government under the Constitution Act of 1852. Continuing immigration created fresh demands for land to be made available, culminating in the New Zealand wars of the 1860s. By 1877 recognition of the Treaty had declined to the point where Chief Justice Prendergast was able to say that the Treaty was a 'simple nullity'. Only in 1975, did the Treaty of Waitangi Act establish the Treaty as having validity in law. This Act also established the Waitangi Tribunal to investigate Maori claims against the Crown (amended in 1985 to permit claims dating back to February 1840).

Rules about the status of the Treaty are the same as English common law rules about the status of treaties in general. This means that the Treaty cannot be enforced domestically unless given effect through statute (Boast, 1992). As a consequence, in recent years most statutes affecting environmental law have been amended to allow recognition of the principles of the Treaty. There is considerable variation in the precise manner in which this is done. One of the strongest statements in law concerning Treaty principles is in the State-Owned Enterprises Act 1986, where section 9 states that, 'nothing in this Act shall permit the Crown to act in a manner that is inconsistent with the principles of the Treaty of Waitangi'. Other resource legislation refers to the Treaty by a variety of formulae: Conservation Act 1987 (section 4: '... to give effect to ...'), Crown Minerals Act 1991 (section 4: '... shall have regard to ...'), and Resource Management Act 1991 (section 8: '... take into account ...'). The RMA also requires local authorities, when preparing or changing regional and district plans, to 'consult' with *tangata whenua*, and 'have regard to' relevant planning documents of a Maori *iwi* authority.

Although guidelines for the application of the Treaty have been published by several Government agencies, such guidelines are not legally binding. However, in July 1989, the then Government of New Zealand released to the public a series of principles by which it would act when dealing with issues arising from the Treaty (*Christchurch Press*, 1989).

- Principle 1: The Principle of Government – The Kawanatanga Principle. The Government has the right to govern and to make laws.
- Principle 2: The Principle of Self Management – the Rangatiratanga Principle. The *iwi* have the right to organise as *iwi* and, under the law, to control the resources they own.
- Principle 3: The Principle of Equality. All New Zealanders are equal under the law.
- Principle 4: The Principle of Reasonable Co-operation. Both the Government and the *iwi* are obliged to accord each other reasonable co-operation on major issues of common concern.
- Principle 5: The Principle of Redress. The Government is responsible for providing effective processes for the resolution of grievances in the expectation that reconciliation can occur.

A major issue is that of *rangatiratanga* (often translated as 'authority' but sometimes as 'sovereignty') over resources and *taonga* (including language and culture). The Waitangi Tribunal (1988) took the position that conservation laws of general application such as the Marine Mammals Act (1978) were a valid application of the Crown's *kawanatanga* (authority to make laws) under Article I of the Treaty, but must also take account of *rangatiratanga* reserved in Article II. As summarised by Boast (1992), 'Maori should be left alone to manage their own resources in their own way, but if Maori self-management threatens the overriding objective of conservation, then it is acceptable – in terms of the Treaty for the Crown to intervene'.

The fundamental importance of the Treaty is that it provides the basis for a legally recognised partnership between Maori and later immigrants in New Zealand. In the words of the New Zealand 1990 Commission, 'today the Treaty continues as a "living document"

– a focus for all New Zealanders to consider its on-going role for our nation and in the partnership between our cultures – today and in the future'.

RMA, the treaty and ethics

As the principal resource management and planning legislation for New Zealand the RMA must fulfil two general criteria. It should achieve its primary aims efficiently and with equity. It should also reflect the underlying *ethos* and aspirations of the people to which it will apply. However, a third requirement is also important for New Zealand – that the Act should honour the principles established in 1840 as a basis for bicultural partnership.

Murray and Swaffield (1994) argue that the RMA is based on four policy myths, using myth in the sense of being, 'a usually traditional story of ostensibly historical events that serves to unfold part of the world view of a people or explain a practice, belief, or natural phenomenon'. Their 'mythical' propositions are that:

- RMA should concern itself with natural and physical resources;
- these resources should be managed sustainably;
- sustainable management should integrate conservation and development; and
- sustainable management is achieved through the rational planning of the environmental outcomes of resource use.

Murray and Swaffield (1994) point out that these myths can simplify, 'a complex and contentious reality', especially in making rational decisions between various options. They also suggest that the myths provide a way to incorporate conflicting values of different interest groups into legislative processes. Part of this involves making transparent the existence and nature of the myths underlying the RMA. They acknowledge that the myths themselves stem from a wide range of value judgments, and 'serve many valued social functions, and only become problematic if their nature is misrepresented or disguised'.

Although for many New Zealand people these myths might appear to be novel and a basis for a new paradigm, they are elements common to both the Maori and Judeo-Christian world view. Rather than being new, they constitute a rediscovery of traditions which have been overlooked in the modern industrial and technological development of New Zealand. It is appropriate that the RMA provides potential for development of a bicultural environmental ethic incorporating strands from the European world view but most importantly incorporating a Maori perspective within a traditional European legal framework.

A new paradigm for significance

The term 'significance' appears in at least 15 places in the RMA but as with several other key words, no definition is offered. There has been a tendency for Europeans in particular to identify *significance* with *things* and *places* rather than with *effects and processes* or *spiritual values* and *perceptions*. Detailed analysis of the meanings of *significance*, including consideration of case law and the recent formulation of New Zealand's national Coastal Policy Statement (Given, 1994b), shows that the term has a wide range of possible meanings including:

- an object which one has in view, expressing itself with weight, influence or high rank;

- an expressive image or graphic representation;
- making known something beyond the object itself and facilitating the understanding and meaning of an abstraction.

In terms of resource management this means taking account of a range of significance types including: *intrinsic significance* (e.g. particular places and things such as mountains, forests and animal and plant species), *representational significance* (e.g. historical landscapes and places), *transcendent significance* (e.g. spiritual issues), and *indirect significance* (e.g. processes such as floods, succession and erosion). Fundamental to *significance* is that it is a *signum* or *sign.*

Significance may lie anywhere on a scale which extends from a point where it is of concern to few people, to a point where it is of paramount concern to a great many people. Thus, a problem in defining parameters for assessing significance is that there are no absolutes. Whereas, one may set absolute limits for various grades of size, shape, weight, and strength, different levels of significance depend on human perceptions. Moreover, the Act requires recognition of significance at national, regional and local (district) levels, and for each of these levels significance may be expressed rather differently. At local level it might be best expressed in terms of specific sites, but at the regional level it is probably more appropriately expressed in terms of landscape and ecosystem. At the national level, significance will be in terms of yet to be determined 'state of the nation' indicators.

Such a wider view of significance accords with both Maori and Judeo-Christian world views. Nothing is regarded as merely an isolated object to be considered only for what it is in physical terms. Kinship obligations (*whanaungatanga* and *manaakitanga*) require that one looks for other meanings – significance lies in historical, spiritual and emotional aspects as well. Consideration of these factors then encourages development of an attitude of reciprocal regard expressed as *aroha.*

Processes of consultation

A departure from former legislation, is that the RMA allows for a very wide range of consultation with the community and interested parties through district and regional planning and the resource consent process (e.g. sections 40, 93, 96), and a wide range of matters must be considered (e.g. sections 6–8, 30, 31, 35, 104, 131). Although the Act preserves the adversary approach to dispute settlement it also embodies provisions for *pre-hearing meetings* for the purposes of, 'clarifying, mediating, or facilitating resolution of any matter or issue' (section 99).

Such provisions accord with the Maori perspective that there is no single world view which should predominate but rather a series of world views which are equally valid. The resolution of a particular resource issue should take account of all views regardless of numerical, social or economic superiority. Resolution should also take account of both *tribal* traditions which have their roots in the distant past, and *living* traditions which continue to evolve through time, regardless of whether they involve Maori or European. Wide consultation, validated by the legal process, facilitates the establishment of *mana* and hence the wider community assumes a voluntary role as *kaitiaki,* taking on a responsibility to care for the environment.

The importance of adequate and *correct* dialogue are well demonstrated in debate over the allocation of New Zealand fisheries. Current reforms of New Zealand fisheries management incorporate the concept of property rights expressed as individual transferable quotas designed along text book lines for control of an 'open access resource'. However, Maori have repeatedly asserted that the Treaty principles give full management authority (*tino rangatiratanga*) over the 'resource' and have made it clear to the Crown that it is not empowered to allocate quota as if 'no one' owned it (O'Connor, 1994).

Wide consultation and a consideration of world views held by Maori people also requires an approach which is not prescriptive at national level. Rather, it should result in establishment of national principles, from which local and community prescriptions can evolve by dialogue, taking into account both the diversity of local contexts and viewpoints, and the individuality of small communities.

A holistic and ecocentric use of resources

Enactment of the RMA heralded a major change of direction in use of resources in New Zealand. A key feature is the focus placed on the *effects* of activities rather than the activities themselves. As a corollary, all the environmental effects of undertaking an activity must be assessed, with the cost of such effects being borne by the developer – the concept that 'the user or polluter pays' (Milne, 1992). *Effects* are defined very widely in section 3 of the Act to include:

(a) Any positive or adverse effect; and
(b) Any temporary or permanent effect; and
(c) Any past, present or future effect; and
(d) Any cumulative effect which arises over time or in combination with other effects – regardless of the scale, intensity, duration or frequency of the effect, and also includes –
(e) Any potential effect of high probability; and
(f) Any potential effect of low probability which has high potential impact.

Similarly, *environment* is defined very widely in section 2, to include almost anything which can be affected by an activity including ecosystems and their parts, natural and physical resources, amenity values, people, and the social, economic, aesthetic and cultural conditions which affect other parts of the environment. Intrinsic values of parts of the environment are given statutory recognition.

The holistic approach of the RMA accords closely with the kinship obligations of *whanaungatanga* and *manaakitanga*, the notion of reciprocity implicit in *utu*, and the implied unconditional responsibilities of care and protection (Patterson, 1994). However, there is ongoing debate over the interpretation of the key concept of 'sustainable management' which is stated in section 5 to be the purpose of the Act (Milne, 1992). The full definition of sustainable management (section 5(2)) has been given earlier and has generated considerable debate which has yet to be resolved. In legislation, the use of the word *while* is unusual; it might well have been expected here to be expressed as *provided that*, which would have established that sustaining, safeguarding and mitigating had clear pre-eminence over 'use, development and protection'. This has been interpreted by some as implying a balance between resource use and conservation, while others have suggested that one can only 'use, develop and protect' so long as one is also sustaining, safeguarding and mitigating at the same time.

The question of a balance between resource use and conservation has other implications (Milne, 1992). Subsection 5(c) of the RMA in requiring that there is, 'avoiding, remedying or mitigating' adverse effects also implies that there will be necessary environmental degradation associated with sustainable management. Another balancing test in section 5 concerns the needs of current generations as against those of the future. No clear guidelines have yet emerged as to the relative merits of present and future generations in allocating resources.

As Milne (1992) points out, 'the start line has only just been reached, Case law will now have to expand and define the notion of sustainability. There are many more questions than answers'. But achieving a consensus on the interpretation of *sustainable development* is not just a question for litigation. The requirement for the RMA to take into account the principles of the Treaty of Waitangi, opens up opportunities for dialogue between Maori and European in coming to that consensus. As pointed out earlier, the concept is implicit in the Maori world view which seeks the appropriate balance between polar pairs, with the view that, firstly, no one perspective is valid at the expense of any other, and secondly, there is a fundamental difference between basic needs and mere wants (Patterson, 1994).

Conclusions

Consideration of the Treaty of Waitangi principles is implicit in contemporary New Zealand legislation including the Resource Management Act 1991 (RMA). Until recently, matters raised by the Treaty have been often regarded as 'add-ons' to the traditional legislative and legal system. The RMA provides opportunity for integrative interpretation of ethical aspects of resource use, especially interpretation of *significance*, methods of consultation with the wider community, and the concept of *sustainable development*. However, such discussion should take place in a forum and by means which are appropriate to both European and Maori culture.

Legislation has to be a significant element in the development of a multicultural environmental ethic for New Zealand. Incorporation of the principles of the Treaty of Waitangi is an important part in that process. The Resource Management Act is itself a *signum* or sign of significance in development of such an ethic, and an important part of its development should be assimilation of the Maori world view expressed in those principles.

It can be argued that this is a necessity and not a choice. O'Connor (1994) quotes the opinion of lawyer Paul Temm (1990), that, 'there is room perhaps for the view that Parliament has its authority by virtue of British sovereignty and British sovereignty has its authority [in New Zealand] by virtue of the Treaty of Waitangi'. As O'Connor suggests in relation to the fisheries resource, 'it is not a question here of what the law of today might say: the law comes *after* the treaty'.

The Resource Management Act provides a legislative framework for sustainable resource use and ultimately for conservation of New Zealand's biodiversity. The Treaty of Waitangi provides a pact of partnership between two cultures and a framework for dialogue (New Zealand 1990 Commission): 'Today the Treaty continues as a "living document" – a focus for all New Zealanders to consider its on-going role for our nation and in the partnership between our cultures – today and in the future'. The Waitangi Tribunal itself has suggested that the Treaty contains certain guarantees of environmental quality. As Boast (1992) points out:

> the notion that the Treaty of Waitangi is itself a guarantor and a symbol of environmental quality is constructive, and is something that should enable everyone to think creatively about the many issues concerning the management and ownership of resources which affect us all.

Ultimately, legislation alone cannot determine how people think or the value systems they will adopt. Achievement of an ethic of sustainable use of biodiversity, of 'partnership' with nature, and of kinship obligations to nature will only come about because people desire such an ethic. Individuals – and so society as a whole – will only want this when convinced that biological diversity has a range of values which encompasses both intrinsic worth and the needs of people, present and future.

New Zealand has already seen landscape transformations which have profoundly affected its unique biological diversity. It belongs to a part of the world – the Pacific region – which at the present time sees much of its biodiversity under threat. Development of a consistent biodiversity ethic to take account of a range of multicultural perspectives will be a key step in stewardship of the unique assemblages of animals and plants in this part of the world.

Acknowledgments

In part this paper results from contracts undertaken for the South Pacific Regional Environment Programme, and for the Canterbury Regional Council, and I am grateful to the administrations of these organizations for releasing relevant portions of contract reports, and for discussion with staff members. These issues have been debated and discussed with a number of colleagues and I am especially grateful to Murray Parsons, Jonet Ward, Nigel Cooper and Bob Carling for their assistance and criticisms.

References

Atkinson, I.A.E. (1985) The spread of commensal species of *Rattus* to oceanic islands and their effects on island avifauna. In *Conservation of Island Birds* (P.J. Moors, ed.) pp. 35–81. Cambridge: International Council for Bird Preservation, Technical Publications No. 3.

Boast, R. (1992) Treaty of Waitangi and environmental law. In *Handbook of Environmental Law* (C.D.A. Milne, ed.) pp. 246–53. Wellington: Royal Forest and Bird Protection Society of New Zealand.

Dahl, A.L. (1986) *Review of The Protected Areas System in Oceania.* Gland, Switzerland: IUCN/UNEP.

Dodson, J.R., ed. (1992) *The Native Lands – Prehistory and Environmental Change in The South Pacific.* Melbourne: Longman Cheshire.

Flannery, T. (1989) Plague in the Pacific. *Aus. Nat. Hist.* **23**, 20–8.

Flannery, T. and Calaby, J. (1990) *Survey of The Mammals of The Solomon Islands.* Report to the Government of the Solomon Islands.

Flenley, J.R., King, A.S.M., Jackson, J., Chew, C., Teller, J.T. and Prentice, M.E. (1991) The late Quaternary vegetational and climatic history of Easter Island. *J. Quatern. Sci.* **6**, 85–115.

Given, D.R. (1980) *Rare and Endangered Plants of New Zealand.* Wellington: Reeds.

Given, D.R. (1991) Biblical aspects of conservation – rediscovering the greenness in Christianity. *Christian Brethren Research Fellowship Journal* **124**, 23–32.

Given, D.R. (1994a) Aspirations, biodiversity and customary rights in the South Pacific – an ABC of problems and solutions for the future? In *Biodiversity and Terrestrial Ecosystems* (C.-I. Peng and C.-H. Chou, eds.) pp. 157–68. Taipei: Institute of Botany, Academia Sinica, Monograph Series 14.

Given, D.R. (1994b) *The Regional Concept of Significance in Relation to the Resource Management Act (1991), Vols. 1 and 2.* Contract Report to Canterbury Regional Council, Christchurch, New Zealand.

Glen. R., ed. (1992) *Mission and Moko.* Christchurch: Latimer Fellowship of New Zealand.
Hight, J. and Straubel, C.R. (1957) *A History of Canterbury, Volume 1.* Christchurch: Whitcombe and Tombs.
Howell, J., ed. (1986) *Environmental Ethics – a New Zealand Contribution.* Christchurch: Lincoln University and University of Canterbury, Centre for Resource Management Special Publication No. 3.
Johnson, T.H. and Stattersfield, A.J. (1990) A global review of island endemic birds. *Ibis* **132**, 169–79.
King, (1990) *Handbook of New Zealand mammals.* Auckland: Oxford University Press.
Laurenson, G.I. (1972) *Te hari Weteriana.* Auckland: The Wesley Historical Society of New Zealand.
McNab, R. (1907) *Murihiku.* Invercargill, New Zealand: William Smith.
Milne, C.D.A. (1992) Resource Management Act 1991. In *Handbook of Environmental Law* (C.D.A. Milne, ed.) pp. 32–95. Wellington: Royal Forest and Bird Protection Society of New Zealand.
Murray, J. and Swaffield, S. (1994) Myths for environmental management. A review of the Resource Management Act 1991. *New Zealand Geographer* **50**, 48–52.
Murray, J.E., Johnson, M.S. and Clarke, B. (1988). The extinction of *Partula* on Moorea. *Pac. Sci.* **42**, 150–2.
New Zealand 1990 Commission (1990) *The Treaty of Waitangi.* Wellington: New Zealand.
O'Connor, M. (1994) Valuing fish in Aotearoa: the Treaty, the market, and the intrinsic value of the trout. *Environ. Values* **3**, 245–65.
Parkin, C. (1986) The human element. In *Environmental Ethics – a New Zealand Contribution* (J. Howell, ed.) pp. 141–68. Christchurch: Lincoln University and University of Canterbury, Centre for Resource Management Special Publication No. 3.
Patterson, J. (1994) Maori environmental virtues. *Environ. Ethics* **16**, 397–409.
Pimm, S.L., Moulton, M.P. and Justice, L.J. (1995) Bird extinctions in the central Pacific. In *Extinction Rates* (J.H. Lawton, and R.M. May, eds) pp. 75–87. Oxford: Oxford University Press.
Potts, T.H. (1882) *Out in the open.* Christchurch: Lyttelton Times.
Pregill, G.K. (1988) Prehistoric extinction of giant iguanas in Tonga. *Copeia* **2**, 505–8.
Robertson, C.J.R., ed. (1985) *Reader's Digest Complete Book of New Zealand Birds.* Sydney: Reader's Digest.
Stevens, G., McGlone, M. and McCulloch, B. (1988) *Prehistoric New Zealand.* Auckland: Heinemann Reed
Temm, P. (1990) *The Waitangi Tribunal: The Conscience of The Nation.* Auckland: Random Century.
Waitangi Tribunal (1988) *Muriwhenua Fishing Report. Wai-22, June 1988.* Wellington: Waitangi Tribunal.
Wardle, P. (1991) *Vegetation of New Zealand.* Cambridge: Cambridge University Press.
Webb, C.J., Sykes, W.R. and Garnock Jones, P.J. (1989) *Flora of New Zealand. Volume IV.* Christchurch: Botany Division, New Zealand Department of Scientific and Industrial Research.
Whistler, A. (1991) Polynesian plant introductions. In *Islands, Plants and Polynesians.* (P.A. Cox and S.A. Banack, eds.) pp. 41–66. Portland, Oregon: Diascorides Press.

Appendix A: The Treaty of Waitangi

Preamble, official English version

Her Majesty Victoria Queen of the United Kingdom of Great Britain and Ireland regarding with Her Royal Favor the Native Chiefs and Tribes of New Zealand and anxious to protect their just Rights and Property and to secure to them the enjoyment of Peace and Good Order has deemed it necessary in consequence of the great number of Her Majesty's Subjects who have already settled in New Zealand and the rapid extension of Emigration both from Europe and Australia which is still in progress to constitute and appoint a functionary properly authorised to treat with the Aborigines of New Zealand for the recognition of Her Majesty's sovereign authority over the whole or any part of those islands.

Her Majesty therefore being desirous to establish a settled form of Civil Government with a view to avert the evil consequences which must result from the absence of the necessary Laws and Institutions alike to the Native population and to Her subjects has been graciously pleased to empower and authorise me WILLIAM HOBSON a Captain in Her Majesty's Royal Navy Consul and Lieutenant-Governor of such parts of New Zealand as may be or hereafter shall be ceded to Her Majesty to invite the Confederated and Independent Chiefs of New Zealand to concur in the following Articles and Conditions.

English version of the Treaty's three articles

Article the first

The Chiefs of the Confederation of the United Tribes of New Zealand and the separate and independent Chiefs who have not become members of the Confederation cede to Her Majesty the Queen of England absolutely and without reservation all the rights and powers of Sovereignty which the said Confederation or Individual Chiefs respectively exercise or possess, or may be supposed to exercise or to possess over their respective Territories as the sole sovereigns thereof.

Article the second

Her Majesty the Queen of England confirms and guarantees to the Chiefs and Tribes of New Zealand and to the respective families and individuals thereof the full exclusive and undisturbed possession of the Lands and Estates Forests Fisheries and other properties which they may collectively or individually possess so long as it is their wish and desire to retain the same in their possession; but the Chiefs of the United Tribes and the individual Chiefs yield to Her Majesty the exclusive right of Pre-emption over such lands as the proprietors thereof may be disposed to alienate at such prices as may be agreed upon between the respective Proprietors and persons appointed by Her Majesty to treat with them in that behalf.

Article the third

In consideration thereof Her Majesty the Queen of England extends to the Natives of New Zealand Her royal protection and imparts to them all the Rights and Privileges of British Subjects.

Maori version of the three articles

Ko te tuatahi

Ko nga Rangatira o te wakaminenga me nga Rangatira katoa hoki ki hai i uru ki taua wakaminenga ka tuku rawa atu ki te Kuini o Ingarangi ake tonu atu te Kawanatanga katoa o o ratou wenua.

Ko te tuarua

Ko te Kuini o Ingarangi ka wakarite ka wakaae ki nga Rangatira ki nga hapu ki nga tangata katoa o Nu Tirani te tino Rangatiratanga o o ratou wenua o ratou kainga me o ratou taonga katoa. Otiia ko nga Rangatira o te wakaminenga me nga Rangatira katoa atu ka tuku ki te Kuini te hokonga o era wahi wenua e pai ai te tangata nona te wenua – ki te ritenga o te utu e wakaritea ai e ratou ko te kai hoko e meatia nei e te Kuini hei kai hoko mona.

Ko te tuatoru

Hei wakaritenga mai hoki tenei mo te wakaaetanga ki te Kawanatanga o te Kuini – Ka tiakina e te Kuini o Ingarangi nga tangata maori katoa o Nu Tirani ka tukua ki a ratou nga tikanga katoa rite tahi ki ana mea ki nga tangata o Ingarangi.

Translation of Maori version – Professor Sir Hugh Kawharu (New Zealand 1990 Commission)

The first

The Chiefs of the Confederation and all the Chiefs who have not joined that Confederation give absolutely to the Queen of England for ever the complete government over their land.

The second

The Queen of England agrees to protect the Chiefs, the Subtribes and all the people of New Zealand in the unqualified exercise of their chieftainship over their lands, villages and all their treasures. But on the other hand the Chiefs of the Confederation and all the Chiefs will sell land to the Queen at a price agreed to by the person owning it and by the person buying it (the latter being) appointed by the Queen as her purchase agent.

The third

For this agreed arrangement therefore concerning the Government of the Queen, the Queen of England will protect all the ordinary people of New Zealand (i.e. the Maori) and will give them the same rights and duties of citizenship as the people of England

10

Collaborative research and intellectual property rights

DARRELL A. POSEY, GRAHAM DUTFIELD and KRISTINA PLENDERLEITH
Working Group on Traditional Resource Rights, Oxford Centre for the Environment, Ethics and Society, Mansfield College, University of Oxford, Oxford, UK

Scientists find themselves working more and more with indigenous, traditional and local communities in all aspects of their collections and investigations. Indigenous knowledge has become increasingly important in research, while at the same time local communities have become increasingly politicized in the use, misappropriation, and commercialisation of their knowledge and biogenetic resources. It is becoming more and more difficult for even the most well-intentioned scientists to stride into indigenous areas and collect plants, animals, folk tales, and photos without having first to convince local leaders that the scholarly efforts will somehow benefit the communities – that the benefits of research results will directly and indirectly lead to strengthening the traditional society. In many parts of the world, indigenous peoples only allow Collaborative Research in which the scientific priorities and agendas are controlled by the communities, or Community Controlled Research in which the communities actually contract scientists to carry out the group's research plan. Control over data has become one of the key 'battle cries' for the indigenous movement, that is now demanding Intellectual Property Rights over information obtained through research and just compensation for economic benefits that eventually may accrue. This paper deals with some of the ethical and practical issues that frame this rapidly evolving debate.

Keywords: Intellectual Property Rights; Collaborative Research; indigenous knowledge; Traditional Resource Rights; codes of ethics

The value and importance of traditional resources

Indigenous knowledge has been defined as 'a cumulative body of knowledge and beliefs handed down through generations by cultural transmission about the relationship of living beings, (including humans) with one another and with their environment' (Gadgil *et al.*, 1993). Traditional resources include plants, animals, and other material objects, which may have intangible (e.g. sacred, ceremonial, heritage, or aesthetic) qualities. Traditional resources may also be totally intangible, metaphysical, or non-quantifiable with no physical manifestations (Posey and Dutfield, in press). Knowledge and traditional resources are central to the maintenance of identity for indigenous peoples; therefore, the control over these resources is of central concern in their struggle for self-determination.

Importance to science

The science of ecology has much to gain from wider access to indigenous knowledge systems. We accept that indigenous peoples can be healers and expert hunters and gatherers, but fewer people realise that they can also be biologists, economic botanists, taxonomists and ecologists.

The pharmaceutical industry continues to investigate (and confirm) the efficacy of many medicines and toxins used by indigenous peoples. Modern agricultural practices, which simplify ecosystems and diminish the genetic diversity of crop species, depend on the free flow of crop germplasm, mainly landraces (sometimes known as folk varieties) bred and conserved by traditional communities over millennia. The use of such varieties by plant breeders makes possible the production of modern varieties containing desirable traits like high productivity and enhanced resistance to certain diseases.

Value to commerce

It has not yet proved possible to estimate the market value of traditional knowledge and its importance for science, but it is certainly enormous, and is bound to increase as advances in biotechnology broaden the range of life-forms containing attributes with commercial applications. For medicines alone, it was estimated that the market value of plant-based medicines (many of which were used first by indigenous peoples) sold in developed countries totalled $43 billion in 1985 alone (Principe, 1989). This figure is often quoted and is not reliable, but whatever the true figure clearly a minimal proportion of it (much less than 0.001%) has ever been returned to the source communities. Companies that produce seeds and agrochemicals, some of which may first have been used by traditional communities, benefit substantially, as do modern farmers. In this way, indigenous and other traditional cultivators subsidize modern agriculture yet obtain no payment in return. Other industries manufacturing personal care products, foods and industrial oils also benefit from the knowledge and resources of indigenous peoples. However, few companies making such products have shown concern for the fact that traditional knowledge is sometimes lost and resources disappear when land is converted, sometimes to produce more of the raw materials for these same companies.

Importance for conservation

Indigenous peoples are already playing vital roles in conservation. Sacred sites frequently function as ecological reserves and are the result of human knowledge in management and conservation. Some governments have acknowledged the value of indigenous knowledge in formulating national biodiversity conservation strategies. For example, Australia's draft National Strategy for the Conservation of Australia's Biological Diversity calls for the Nation to:

> Recognize the value of the traditional knowledge and practices of Aboriginal people and Torres Strait Islander and integrate this knowledge and those practices into biological diversity research and conservation programmes by:
>
> – encouraging the recording (with the approval and involvement of the indigenous people concerned) of traditional knowledge and practices;
>
> – assessing their potential value for nutritional and medicinal purposes, wildlife and protected area management and other purposes; and
>
> – applying traditional knowledge and practices in ways which ensure the equitable sharing of the benefits arising from their use (The Biological Diversity Committee, 1992, in Sutherland, 1993).

Indigenous views and values

The first concern of indigenous peoples is that their right NOT to sell, commoditize, or have expropriated from them certain domains of knowledge and certain sacred places, plants, animals, and objects be respected. The Kari-Oca Declaration and Indigenous

Peoples' Earth Charter Clause 95 states that 'Indigenous wisdom must be recognized and encouraged', but warns in Clause 99 that 'Usurping of traditional medicines and knowledge from indigenous peoples should be considered a crime against peoples'.

Clause 102 of the Kari-Oka Declaration is explicit about indigenous peoples' concern on IPR issues:

> As creators and carriers of civilizations which have given and continue to share knowledge, experience and values with humanity, we require that our right to intellectual and cultural properties be guaranteed and that the mechanism for each implementation be in favour of our peoples and studied in depth and implemented. This respect must include the right over genetic resources, genebanks, biotechnology and knowledge of biodiversity programmes.

Since the Earth Summit, dozens of conferences, seminars and workshops have been held by indigenous peoples to discuss the evolving Intellectual Property Rights (IPR) debate. For example, in the Charter of the Indigenous-Tribal Peoples of the Tropical Forests (Article 44) IPR concerns are explicitly stated:

> Since we highly value our traditional knowledge and believe that our biotechnologies can make an important contribution to humanity, including 'developed' countries, we demand guaranteed rights to our intellectual property, and control over the development and manipulation of this knowledge.

In 1993, the 'First International Conference on the Cultural and Intellectual Property Rights of Indigenous Peoples' was held in Whakatane, Aotearoa New Zealand. Over 150 Delegates from 14 countries met to prepare 'The Mataatua Declaration on Cultural and Intellectual Property Rights of Indigenous Peoples', which, reflecting the concerns of the delegates, was passed by the Plenary The Declaration states that cultural and intellectual property are central to the right of determination and that, although the knowledge of indigenous peoples is of benefit to all humanity, the first beneficiaries of indigenous knowledge must be the direct indigenous descendants of such knowledge.

The basis for indigenous concern

Why are indigenous peoples so concerned about IPR? Consider these examples:

The Urueu-Wau-Wau, Merck and the tiki uba

In a 1988 issue of *National Geographic Magazine*, Loren McIntyre describes the 'Last Days of Eden' for the 350 members of the Amazonian Urueu-Wau-Wau tribe. They are portrayed as being vulnerable to diseases carried by outsiders and trying to resist the encroachments of settlers on their lands. Three photos on one of the pages, one of which shows a tapir bleeding from an arrow wound, are accompanied by the following caption (McIntyre, 1988):

> Secrets of rain forest chemistry provide a feast for the Urueu-Wau-Wau. Using poison arrows, they down a young tapir that bumbled into their village at night. Wooden arrow points are coated with sap squeezed from the stringy red bark off tiki uba trees and hardened by fire. An anticoagulant, tiki uba causes victims to bleed to death. In addition to such deadly jungle lore, knowledge of potentially useful foods and drugs, accumulated over thousands of years, may be lost forever if the forest and its inhabitants disappear.

Jesco von Puttkamer, who took photographs which accompanied the article, was quoted as saying in reference to the plant: 'I think it may be a great pharmaceutical find.' This article attracted the attention of researchers working for Merck, the world's biggest pharmaceutical company, and von Puttkamer agreed to send them bark and sap specimens in order for them to carry out tests (Jacobs *et al.*, 1990). These tests confirmed that the bark contained at least one compound that inhibited enzymes that cause blood clotting, and might therefore be useful in heart surgery.

One can make the argument that McIntyre and von Puttkamer were acting in the best interests of humankind when they described the tiki uba in their article. However, by doing so they made it possible for a drug company to appropriate their knowledge without any obligations to compensate the Urueu-Wau-Wau, who, in their present precarious situation could perhaps benefit a great deal from compensation. Had Merck continued its research and developed a lucrative product the Urueu-Wau-Wau would have been in a weak position to demand benefits due to them as conservers of the plant and original discoverers of its useful attributes.

Neem– a traditional and 'modern' bio-pesticide

Seeds of a species of neem tree (*Azadirachta indica*) have been ground and scattered on fields by Indian farmers for centuries to protect their crops from insect pests. However, the neem tree has many other uses. To give just a few examples: it appears to be effective against malaria and internal worms; the leaves are used to protect stored grain from pests and clothes from moths; neem oil is used to make candles, soap and a contraceptive, and can even fuel diesel engines; and 500 million Indians reportedly use neem for toothbrushes (Shiva, 1994). These uses were discovered over the millennia by tribal and rural Indian communities.

As a pesticide, neem has a rare ability to target crop pests without harming other organisms. Also, it need not compete for space with crops, making it suitable for use in agroforestry systems. Thus it has great potential as a cheap and environmentally-friendly alternative to commercial synthetic pesticides (van Latum and Gerrits, 1991).

Two companies, W.R. Grace and Agrodyne, recently obtained patents in the US for derivatives of the active principle developed in their laboratories, even though the insecticidal, human non-toxic and biodegradable properties of neem are far from novel and non-obvious to millions of Indian farmers. One patent is for a more stable, easier to store, and thus more useful, synthetic derivative of the naturally occurring chemical, 'azadirachtin' (McGowan, 1991; Axt *et al.*, 1993). The other is for a more stable storage form of azadirachtin modified sufficiently in laboratories to be considered no longer a product of nature. Another patent has been granted in the USA for a pre-treated extract of neem bark effective against certain cancers. W.R. Grace is producing neem-based pesticides with an Indian company called PJ Margo in India (Sharma, 1994). They estimate that the global market for their product may reach $50 million per annum by 2000 (*AgBiotechnology News*, 1993). Agrodyne Technologies now has US government approval to sell neem-based bio-insecticides, and has applied for registration of its products in several European and Latin America countries. Showa Biochemicals is currently marketing two neem-based products in Japan: one is an insect repellent, and the other is a treatment for gastrointestinal problems (RAFI, 1993).

Obviously, these companies stand to gain from the insights of Indian farmers. Nevertheless, they have no legal obligations towards them, even if it were possible to identify individuals or communities more deserving of compensation than others, because the knowledge was in the public domain. India has a weak claim to any compensation because the tree is native also to India's neighbours, and because it is now grown around the world.

One important issue highlighted by these cases is that indigenous knowledge is not treated by industry as being owned. On the other hand, companies and governments can acquire such knowledge from a community directly, or more commonly through literature searches, explore its commercial potential and seek IPR protection (in the form of patents, copyrights, trademarks, etc.). Subsequently, the company may acquire a legal monopoly to manufacture and sell products developed from the application of this knowledge. This pattern is not only parasitic on indigenous peoples, but also on the environment, since knowledge and raw materials are considered to be 'free' and require no restitution.

A call for partnerships

Signatories to the internationally legally binding Convention on Biological Diversity (CBD) have pledged to:

> respect, preserve and maintain knowledge, innovations, and practices of indigenous and local communities embodying traditional lifestyles relevant for the conservation and sustainable use of biological diversity and promote their wider application with the approval and involvement of the holders of such knowledge, innovations and practices and encourage the equitable sharing of the benefits arising from the utilization of such knowledge, innovations and practices (Article 8(j)).

Governments are acknowledging that there is a need for a new relationship between themselves and indigenous peoples and even State agencies have developed ethical codes of conduct for research, such as the Canadian Royal Commission on Aboriginal Peoples (Sutherland, 1993).

What can scientists do to put into practice this call for new partnerships? A rapidly growing number of scientific and professional societies have already developed codes of conduct. Examples include the draft guidelines of the Society of Economic Botany, the 1992 Asian Symposium on Medicinal Plants, Spices and Other Natural Products, which developed a declaration known as the Manila Declaration concerning the Ethical Utilization of Asian Biological Resources, and a code of ethics developed at the Botany 2000 Herbarium Curation Workshop held in 1990 in Australia. Guidelines for ethical collecting for botanical gardens and research institutions have been prepared by Cunningham (1993a). The International Society for Ethnobiology at its Congress in Lucknow in November 1994 decided to initiate a process to develop a code of ethics for ethnobiological research. In addition, the Pew Conservation Scholars have recently been working together to produce ethical guidelines for 'accessing and exploring biodiversity', which would encompass research activity whether or not it is commercial in intent and would embrace the principles of prior informed consent and equitable sharing.

Indigenous peoples have already produced their own collaborative research guidelines for researchers who visit their lands. For example:

The Inuit Tapirisat of Canada

The Inuit Tapirisat of Canada produced a background paper ('Negotiating research relationships in the North') containing a useful list of principles based on existing ethical guidelines and on specific concerns expressed by members of Inuit communities:

(i) Informed consent should be obtained from the community and from any individuals involved in research.
(ii) In seeking informed consent the researcher should at least explain the purpose of the research; sponsors of research; the person in charge; potential benefits and possible problems associated with the research for people and the environment; research methodology; participation of or contact with residents of the community.
(iii) Anonymity and confidentiality must be offered and, if accepted, guaranteed except where this is legally precluded.
(iv) On-going communication of research objectives, methods, findings and interpretation from inception to completion of project should occur.
(v) If, during the research, the community decides the research is unacceptable, the research should be suspended.
(vi) Serious efforts must be made to include local and traditional knowledge in all stages of research including problem identification.
(vii) Research design should endeavour to anticipate and provide meaningful training of aboriginal researchers.
(viii) Research must avoid social disruption.
(ix) Research must respect the privacy, dignity, cultures, traditions and rights of aboriginal people.
(x) Written information should be available in the appropriate language(s).
(xi) The peer review process must be communicated to the communities, and their advice and/or participation sought in the process.
(xii) Aboriginal people should have access to research data, not just receive summaries and research reports. The extent of data accessibility that participants/communities can expect should be clearly stated and agreed upon as part of any approval process.

The Kuna

The Kuna are an indigenous group who inhabit a coastal area of Panama that has achieved a degree of autonomy that is uncommon in Latin America. In 1988 two Kuna institutions, the Proyecto de Estudio para el Manejo de Areas Silvestres de Kuna Yala (PEMASKY) and the Asociacion de Empleados Kunas (AEK) in Panama produced a manual for researchers entitled 'Research Programme: Scientific Monitoring and Cooperation' (Programa de Investigacion Monitoreo y Cooperacion Cientifica). This document reflects the interest of the Kuna in integrating their traditional ecological knowledge and western science through Collaborative Research based on priorities set by the Kuna. The aim is to generate information for the benefit of the Kuna themselves in the following priority areas: ecological research, botanical and faunal inventories, soil surveys, socioeconomic studies, ethnobotanical studies, and the study and recording of Kuna traditions and culture. The manual contains regulations for visiting scientists, including descriptions of benefits that should be returned to the Kuna. Thus, researchers are required to:

(i) Develop a proposal outlining the timing, extent and potential environmental and cultural impact of a research programme. This must be approved by the Scientific Committee of PEMASKY.
(ii) Provide PEMASKY with written reports of the research, and two copies of any publications, in Spanish.

(iii) Provide PEMASKY with copies of photographs or slides taken during the research programme.
(iv) Include in their research programme Kuna collaborators, assistants, guides and informants, and should undertake training in relevant scientific techniques.
(v) Provide descriptions of all species new to science.
(vi) Receive approval for the collection of species from the Scientific Committee of PEMASKY. All collections must be done in a non-destructive manner, may not include any endangered species, and may not be used for commercial purposes. Samples of all collected specimens must be left with PEMASKY (to be added to collections at the University of Panama).
(vii) Undergo an orientation into the culture of the Kuna Yala, and must respect the norms of the communities in which they work.

The procedures also forbid the introduction of exotic plant or animal species, or the manipulation of genes. Research is restricted to certain areas of the reserve, is prohibited in some sites, such as ceremonial or sacred sites, and is controlled in other specific sites, such as some forest areas under community management.

The Awa Federation and the New York Botanical Garden

In April 1993 the NYBG signed a research agreement with the Awa Federation who live in the biologically diverse Province of Cachi in Ecuador. The Awa Federation is a legal institution which administers the land held under communal title by the Awa. The Awa have only had legal recognition as citizens and collective title to their land since 1988, but since then have been active in demarcating their territory and expelling colonist settlers. The two-year agreement, known as the 'Reglamentos para la Realizacion de Estudios Cientificos en al Territorio de la Federacion Awa', includes the following regulations:

(i) All scientists must ask for written permission to carry out studies. The written request for permission must include a description of objectives, size and composition of research party, length of research programme, species or object of study, and the manner in which this research will benefit the Awa community.
(ii) The request for permission must be given with a minimum of two months' notice. [This allows time for widely dispersed communities to meet and reach a decision.]
(iii) More than five people to a research group is prohibited.
(iv) Local guides and informants must accompany all scientists.
(v) The removal of any object from Awa territory not approved by the federation is prohibited.
(vi) The prices established by the Awa Federation for their services:
- each member of each scientific group must contribute 1000 sucres in order to enter Awa territory to the Federation;
- guides and informants receive 700 sucres per day;
- cooks, cleaners, and other workers receive 500 sucres per day;
- members of scientific groups from Ecuadorean universities or institutions pay only 500 sucres per day to enter Awa territory.

(vii) The Awa Federation must receive acknowledgement in publications.

As regards putting some of these kinds of ideas and principles into practice, some progress has been made. For example, Sutherland (1993) mentions a successful collaborative research project in the Uluru National Park of Australia which had depended on the ecological knowledge of the Anangu people. It appears to have been a success because:

> . . . the Anangu owned the land on which it was conducted, there had been two-way information flows, Anangu had decision-making power and had been involved in all stages of the project, gender-specific skills were recognized, Anangu experts were paid expert consulting rates, flexible work arrangements and good working relationships were developed, and the Anangu vetted all information before publication (Sutherland, 1993).

In the absence of their own research and documentation facilities, indigenous peoples are most likely to be the 'subjects' of other peoples' research. This increases the ethical imperative of negotiating to conduct field research, rather than simply ignoring any concerns and legitimate interests that local people might have.

Significantly, in recent years there have been cases of indigenous people hiring researchers who agree to the community or tribe holding the copyright. For example communities in the Solomon Islands have secured copyright for their ecological knowledge recorded by researchers. The Arctic Institute of North America in Canada has developed a concept called 'participatory action research' from the idea that researchers should be commissioned to do research, enabling the communities to authorise publication of their cultural information (Greaves, 1993).

Problems in collaboration

Alienation of resources

Academics rarely become rich from recording traditional knowledge, and might find it surprising that others do. Often, however, they may be unaware of the full implications of publications. When the knowledge of a traditional community is published, it immediately falls into the public domain, meaning that it can be used by anybody, including companies that find the knowledge useful and valuable to their R&D programmes. Even though the book or research report was dependent on information provided freely by indigenous people, it is copyrighted by the researcher/writer, publishing company or sponsor of the research. Government or university sponsors often justify holding copyright because public funds were used to support the research project (see Mead, 1993; Ruppert, 1994). For example, a project funded by the European Union to survey the ethnobotany of the Topnaar people of Namibia involved not only the export of medicinal plants by the researchers, but the claim by the European Commission that all research results were their property (Cunningham, 1993b). Although plant samples were left with the national herbarium and research results were passed on to the Namibian authorities, this is more likely to benefit the Namibian government than the very people whose cooperation made the project possible.

Failure to acknowledge indigenous sources is another issue about which some indigenous peoples have complained. Co-authorship with indigenous 'cultural consultants' is now common in anthropology and, to some extent, has alleviated this problem.

Restricting readers through published alerts may be somewhat effective in guaranteeing the proper use of published material. For example, in a recent publication, Elisabetsky and Posey (1994) advise readers that the information contained in their article was authorised and freely given by indigenous leaders. In the paper's opening paragraph, readers are advised that they are morally and ethically bound to respect the source of the information and to share any benefits that might accrue, economic or otherwise, with the indigenous community. Although in some countries such a warning may not have legal force, it nonetheless carries a universal force of moral and ethical standards and obligations (Elisabetsky and Posey, 1994).

Defensive publication may be effective in some cases to put information into the public domain so as to block a patent application from somebody else (The Crucible Group, 1994). The danger is that the applicant may still be able to obtain a patent if he or she can convince the patent examiners that the invention described is in some way an inventive step beyond that described in the defensive publication.

Publishing is not the only way to put traditional resources into the public domain; germplasm collections do the same thing. Many collectors of plants and other biological material are academics with contracts from industry. These contracts make it possible for researchers to continue with their under-funded botanical, pharmacological or other academic research. Sometimes a practical distinction between collections for academic and those for commercial ends is not made clear to the communities with which researchers work. From the indigenous standpoint, all collections alienate their communal property from their rightful holders in the society.

Control and access by local communities

Indigenous peoples are frequently denied the rights to full control over and access to their lands, territories, and the natural resources found within them. The International Covenant on Economic, Social and Cultural Rights (ICESCR) and the International Covenant on Civil and Political Rights (ICCPR), however, guarantee that:

> all peoples may, for their own ends, freely dispose of their natural wealth and resources without prejudice to any obligations arising out of international economic co-operation, based upon the principle of mutual benefit, and international law. In no case may a people be deprived of its own means of subsistence.

Article 15(1c) of the ICESCR provides that

> the States Parties to the present Covenant recognise the right of everyone:
> (c) To benefit from the protection of the moral and material interests resulting from any scientific, literary or artistic production of which he is the author.

The 1989 International Labour Organisation's Convention 169 Concerning Indigenous Peoples in Independent Countries states as follows:

> The peoples concerned shall have the right to decide their own priorities for the process of development as it affects their lives, beliefs, institutions and spiritual well-being and the lands they occupy or otherwise use, and to exercise control, to the extent possible, over their own economic, social and cultural development. In addition, they shall participate in the formulation, implementation and evaluation of plans and programmes for national and regional development which may affect them directly (7.1).
> The rights of the peoples concerned to the natural resources pertaining to their lands shall be specially safeguarded. These rights include the right of these peoples to participate in the use, management and conservation of these resources (15.1).

All these provisions support the view that international law gives to indigenous peoples the right to safeguard their own resources and to benefit from their knowledge and from goods produced or owned by them, whether or not they wish to commercialise them. This is becoming an increasingly important issue, since the CBD clearly assigns all biological and biogenetic resources to States. Indigenous peoples fear that some States will now try to nationalise their knowledge and traditional technologies.

Compensation

The breadth of the IPR debate has outgrown its name, since the legal instruments of intellectual property rights are inadequate to protect the cultural, scientific and intellectual resources of indigenous peoples. The inherent dangers lying within the IPR debate are well recognized by many indigenous peoples, who are highly sceptical of suggestions that they can benefit from IPR as it is understood in the Western sense, and are keen to take the lead and confront the problem on their own terms.

According to Alejandro Argumedo (personal communication, 1994) of the Indigenous Peoples Biodiversity Network, an organisation established to provide a distinctive indigenous voice on conservation and sustainable use of biological diversity:

> ... we are concerned about the way IPR is being presented to our peoples. Particularly we are concerned by the strategy of using IPR to push indigenous communities, especially traditional ones, into embracing values (commoditization of the sacred, short term profit making, etc.) which are highly corrosive of our cultural values... [A]ctivities often referred to as 'beneficial' for our peoples and that have an IPR component such as bioprospecting, biodiversity development projects, extractive reserves, etc. are now a trend and are playing that role.

The term Traditional Resource Rights (TRR) has emerged to define the many 'bundles of rights' that can be utilized for protection, compensation, and conservation. The change reflects an attempt to build upon the concept of IPR protection and compensation, while recognizing that traditional resources – both tangible and intangible – are also covered under a significant number of other international agreements. The term 'property' was dropped, since property for indigenous peoples frequently has intangible, spiritual manifestations, and, although worthy of protection, can belong to no human being. Privatization or commoditization of these are not only foreign, but incomprehensible or even unthinkable. Nonetheless, indigenous and traditional communities are increasingly involved in market economies and, like it or not, are seeing an ever-growing number of their resources traded in those markets.

Conclusion

Ecologists and other scientists are encountering a rapidly changing political landscape for their research. Indigenous peoples are seeking more control over research and collecting undertaken on their lands and territories as well as data that result. The Convention on Biological Diversity (CBD) calls for greater decision-making and participation by local communities in conservation, sustainable development, education and training. The CBD also provides for wider application of traditional knowledge, innovations and practices, together with benefit sharing. Scientists will need to become better informed and prepared to deal with problems of authorisation from and access to indigenous, traditional, and local communities, which will undoubtedly involve more Collaborative Research and Community-Controlled Research. Intellectual Property Rights provisions are becoming mainstream and imply that ecologists and other researchers will have to develop their own guidelines, principles and codes of conduct for collection, data access, publication, and benefit sharing.

References

Axt, J.R., Att, J.R., Corn, M.L., Lee, M. and Ackerman, D.M. (1993) *Biotechnology, Indigenous Peoples, and Intellectual Property Rights*. Washington DC: CRS/The Library of Congress.

The Crucible Group (1994) *People, Plants and Patents*. Ottawa: International Development Research Centre.

Cunningham, A.B. (1993a) *Ethics, Ethnobiological Research, and Biodiversity*. Gland: World Wide Fund for Nature.

Cunningham, A.B. (1993b) *Conservation, Knowledge and New Natural Products Development: Partnership or Piracy*? Paper presented at the conference: Intellectual Property Rights and Indigenous Knowledge, Granlibakken, Lake Tahoe.

Elisabetsky, E. and Posey, D.A. (1994) Ethnopharmacological search for antiviral compounds: treatment of gastrointestinal disorders by Kayapo medical specialists. In *Ethnobotany and The Search For New Drugs (Ciba Foundation Symposium 185)*. (D.J. Chadwick and J. Marsh, eds) pp. 77–94. Chichester: John Wiley & Sons.

Gadgil, M., Berkes, F. and Folke, C. (1993) Indigenous knowledge for biodiversity conservation. *Ambio* **22**, 151–6.

Greaves, T. (1993) *Tribal Rights*. Paper presented at the conference: Intellectual Property Rights and Indigenous Knowledge, Granlibakken, Lake Tahoe.

Jacobs, J.W., Petroski, C. Friedman, P.A. and Simpson, E. (1990) Characterization of the anticoagulant activities from a Brazilian arrow poison. *Thrombosis and Haemostasis* **63**, 31–5.

van Latum, E.B.J. and Gerrits, R. (1991) *Bio-pesticides in Developing Countries: Prospects and Research Priorities*. Maastricht: ACTS Biopolicy Institute.

McGowan, J. (1991) Who is the inventor? *Cultural Survival Quarterly*, **15**(3) 20.

McIntyre, L. (1988) Last days of Eden. *National Geographic*, **174**, 800–17.

Mead, A.T.P. (1993) Delivering good services to the public without compromising the cultural and intellectual property rights of indigenous peoples: the economics of customary knowledge. *New Zealand Institute of Public Administration, Research Papers*, **X**, 31–6.

Posey, D.A. and Dutfield, G.M. assisted by Griffiths, T., Laird, S. and Plenderleith, K. (in press) *Beyond Intellectual Property Rights: Towards Traditional Resource Rights for Indigenous and Local Communities*. Ottawa and Gland: International Development Research Centre and World Wide Fund for Nature.

Principe, P.P. (1989) The economic significance of plants and their constituents as drugs. In *Economic and Medicinal Plants Research, volume 3* (H. Wagner, H. Hikino and N.R. Farnsworth, eds.) pp. 1–17. London, San Diego: Academic Press.

Ruppert, D. (1994) Buying secrets: Federal Government procurement of intellectual cultural property. In *Intellectual Property Rights for Indigenous Peoples: a Sourcebook* (T. Greaves, ed.) pp. 111–28. Oklahoma City: Society for Applied Anthropology: 111–128.

Rural Advancement Foundation International (1993) RAFI Communique. December.

Sharma, D. (1994) *GATT and India: the Politics of Agriculture*. Delhi: Konark Publishers.

Shiva, V. (1994) Freedom for seed (interview with Satish Kumar). *Resurgence* **March/April**, 36–9.

Sutherland, J. (1993) *National Overview of Policies, Protocols and Legislation Dealing With Indigenous Australians' Intellectual and Cultural Property*. Paper presented for Rainforest Aboriginal Network and Wet Tropics Management Authority Workshop, 25–27 November.

11

Ethical considerations and animal welfare in ecological field studies

R.J. PUTMAN

Department of Biology, University of Southampton School of Biological Sciences, Boldrewood Building, Bassett Crescent East, Southampton SO16 7PX, UK

This paper addresses some of the ethical and welfare considerations implicit in the application of general techniques in common use during the course of collecting data in ecological field work. Even if they are not explicitly constructed as manipulative experiments, many field studies involve some degree of intervention during routine monitoring programmes: through disturbance caused merely by the presence of an observer or where specific sampling techniques themselves involve capture, handling and marking. Such interventive techniques may cause discomfort, distress or loss of fitness, even in the extreme may result in incidental mortality – and the ethical scientist should critically evaluate the implications of each methodology before adopting any procedure. The paper reviews by way of example the types of objective information now available for both small and larger mammals in relation to: (i) distress and mortality during capture operations; (ii) mortality or distress caused at the time by marking; (iii) longer-term consequences of handling and marking in terms of subsequent [delayed] mortality or loss of fitness, before considering a formal framework for assessment of costs and benefits of any given field programme.

Keywords: environmental ethics; animal welfare; animal suffering; trapping; animal tagging

Introduction

Research in population or community ecology is concerned primarily with seeking to explain patterns in the abundance or distribution of living organisms, looking for factors which affect the relative abundance of different species and promote or limit their coexistence.

In their search for the 'rules' governing the dynamics of individual animal and plant populations, population ecologists in the early days for the most part withdrew to the laboratory where they could escape the confusing complexity of the wealth of interactions in any natural community and study the effects of individual factors in turn under strictly controlled conditions. Increasingly, however, it became apparent that the results from such contrived experiments might not truly reflect dynamics in the real world of more complex interaction in truly multi-species systems where the effects of competition or predation might be altered through the presence of third parties involved in higher order interaction (e.g. Strauss, 1991) and population ecologists interested in dynamics of population interaction in the real world were forced to examine their dynamics in field situations.

However, quantifying the effects of such processes in the field – even proving that such interactions did have any effect upon the dynamics of the populations concerned – proved fraught with difficulty. The very complexity of the system made distinction of the effects of individual interactions extremely difficult. In addition, in studies of natural systems one is of necessity studying the end product of many years of interaction; presented with the end result it is commonly extremely difficult to attribute definitive cause. Community ecologists were faced with the same problem: while convinced that biotic interactions of various sorts were extremely important in generating the patterns they observed of species co-occurrence and relative abundance, they were by and large presented with a *fait accompli* and commonly could not distinguish which of a variety of alternative factors might have resulted in such arrays.

Recognising the limitations of both laboratory-based population studies and community-analyses based on simple observation of current pattern, ecologists were encouraged to embrace a more experimental approach to their field studies and embark on manipulative experiments to test explicit hypotheses rather than rely on *post hoc* rationalisation of general theory, or 'soft corroboration' of established theories (see e.g. Strong *et al.*, 1984). Through the late 1960s up till the present this recognition that our ecological understanding could only advance through more formal experimentation has led to an increasing number of manipulative field experiments – in exploring, e.g., the importance and role of competition as a process at population or community level, or the idea of keystone species within community webs. Formal rigorous protocols have become established (e.g. Bender *et al.*, 1984; Wiens, 1989) and the literature witnesses a proliferation of such manipulative experiments.

But despite their scientific 'power' such field experimentation in removal of key species from some natural system or artificial manipulation of relative abundances clearly have some ethical implications: what, we might ask actually happened to the individual *Ptethodon* or *Pisaster* removed in the classic experiments of Hairston (1980) or Paine (1966, 1969)? And what are the moral issues raised by removal experiments involving keystone species which, if 'successful', often result in dramatic repercussions throughout the rest of the community of dependent organisms? On the other hand, given that as rigorous scientists we may accept the need for experimentation if we are ever to gain an understanding of the rules and mechanisms governing the ecological patterns we observe, do we have any realistic alternatives? Opportunities to 'cash in' upon past 'experiments' (historical interventions contrived 'before we knew any better') are perhaps few and far between (although Diamond, Pimm and co-workers are to be commended for their continued exploitation of such past misdemeanours: Crowell and Pimm, 1976; Moulton and Pimm, 1983, 1986; Diamond *et al.*, 1989) and other workers might be exhorted to do the same (Diamond, 1983). At least such prior 'experiments' have the merit that one does not have to wait a further 50 years for the result!

Such deliberate manipulative experiments in ecology are, however, something of a special case, are very explicit and overt instances of clear intervention and as such are immediately obvious candidates for ethical debate. A very useful review of the issues raised is presented by Farnsworth and Rosovsky (1993). In this paper I intend to focus upon the more widespread, but less explicit intervention implicit in more general ecological fieldwork, in the application of general *techniques* often in common every-day use during the course of collecting basic data in a variety of population, community or behavioural ecology studies.

Even if they are not explicitly constructed as manipulative *experiments*, many ecological field studies involve some degree of intervention during routine monitoring programmes: through disturbance caused merely by the presence of an observer or where specific sampling techniques themselves involve capture, handling and marking. Such interventive techniques may cause discomfort, distress or loss of fitness, even in the extreme may result in incidental mortality – and the ethical scientist should critically evaluate the implications of each methodology before adopting any procedure – even in the limit abandoning the research proposal if he/she cannot adequately justify its value in relation to the welfare costs attached.

Objective information is readily available for a lot of these techniques for many species; thus decisions can often be taken against a background of hard data and there is no easy 'excuse' for not addressing these issues formally in relation to one's own work. In the rest of this paper therefore I will attempt to illustrate some of the types, and sources of information now available for various vertebrates in relation to (i) distress and mortality during capture operations; (ii) mortality or distress caused at the time by marking; (iii) longer term consequences of handling and marking in terms of subsequent (delayed) mortality or loss of fitness (e.g. implications among social species of wearing a collar/eartag), before suggesting a 'decision-making' framework of questions which ethical scientists might wish to ask themselves before embarking on any programme of interventive fieldwork.

Legislative constraints and cost-benefit analyses

Many of the techniques employed in such routine field study are of course covered by restrictive legislation and need special licence (e.g. invasive procedures, such as blood sampling, are explicitly covered in the UK by Home Office regulations; many forms of live-trapping/live capture need special licence of exemption under the Wildlife and Countryside Acts or other relevant legislation). Within the UK, the main legislative orders affecting ecological and behavioural field work are the provisions of the Wildlife and Countryside Act 1981 (and as amended 1985, 1991); where work involves administration of anaesthetic or procedures which may cause pain or suffering or result in long-term damage, such procedures also require licensing under the Protection of Animals (Anaesthetics) Acts 1954, 1964 and subsequent Animals (Scientific Procedures) Act 1986. I should stress that these are by no means the only Acts containing clauses relevant to the field ecologist and it is beyond the scope – and the intention – of this paper to provide a comprehensive or exhaustive review of all relevant legislation. In my own research on deer for example, I come under specific provision of the Deer Acts 1967 and 1991 and individual researchers should carefully check through all relevant legislation affecting their study systems or intended procedures. Useful summaries of current legislation affecting British wildlife are provided by e.g. Cooper (1987, 1991).

Licensing authorities have considerable discretion in whether or not to issue a licence in the first place and considerable room for manoeuvre to enforce welfare issues through the imposition of special conditions. The Wildlife and Countryside Act and its amendments protect all wild plants which occur naturally in the wild in Great Britain (with some exceptions under specific, defined, circumstances); in addition the Act protects all birds (with similar exception), the majority of land vertebrates (including amphibians), some fish and sea mammals. Taking or killing of protected creatures for scientific or educational purposes may be permitted under licence. Such restriction under the Wildlife and Countryside Acts applies *inter alia* to capture of deer (protected also under the Deer Acts), bats, shrews and to mist-netting of birds or ringing birds at the nest.

In granting licence, the appropriate authority (English Nature, EN, Countryside Commission for Wales, CCW, Scottish Natural Heritage, SNH) may make the licence general or specific to any degree and may choose to impose special conditions. Licences are generally issued only where the applicant has proven experience in capture and handling of the relevant species and where the scientific case is strong, although as noted, licences may also be granted on purely educational grounds.

Costs – in terms of animal life and suffering – and benefits are also explicitly considered before a project licence may be granted by the Home Office under the Animals (Scientific Procedures) Act: where the calculated or incidental loss of life and level of short and longer term suffering are weighed against the perceived benefits of the research for animal or human life. Here too the experience of the individual researcher as well as the implications of the procedure proposed are considered carefully before licence may be granted. In all cases, in reaching their decision about what licence to grant – and under what conditions, the Home Office, or other statutory body (EN, CCW, SNH etc.) consider the experience and 'attitude' of the researcher and explicitly assess the likely costs and benefits of the proposed research.

By no means all research procedures need Home Office licence, or licensing under the various wildlife Acts; in any case, simply relying on someone else's decisions on whether or not to grant you a licence to determine your ethical position is simply passing the buck. Some professional organisations and societies now issue guidelines to their members (e.g. Association for the Study of Animal Behaviour 1981); a number of journals now refuse to publish papers where data appear to have been collected by methods/protocols considered unethical, but again we should not allow restrictions in law or guidelines proffered by professional institutes to relieve us of the need to consider the ethics of our own activities; it is up to each individual to determine whether or not procedures he/she proposes to adopt are justifiable to themselves. However, the factors now considered by the statutory authorities in determining and controlling the granting of such licences establish at least evaluation systems by which we might assess other, unrestricted activities.

Nor should it be thought that such self-scrutiny need be restricted to research activity alone. While I focus attention here primarily on issues raised at research level, many of the same problems also apply to educational exercises such as school or college field studies. Particularly in the case of invertebrate sampling, destructive sampling methods are often employed; further, some incidental mortality to target or non-target species is inevitable even in the use of non-destructive techniques of sampling such as live-trapping of small mammals (see below). Professional researchers may be becoming increasingly aware of the ethical implications of their own research procedures but how many small mammals and invertebrates die each year – and unconsidered – in the cause of 'education' in student field courses?

Assessing 'costs' and 'benefits'; value judgements and objective data

Although some possible approaches to weighing costs in animal suffering against research benefit are explored by, for example Smith and Boyd (1991), in any such cost/benefit analysis both 'costs' and 'values' to an extent must be subjectively determined. Perceived benefits and costs also depend on whether one adheres to a *scientistic*, *utilitarian*, or *moralistic* code of ethics (Kellert, 1991; Farnsworth and Rosovsky, 1993), or, to whether one (i) believes in the pursuit of science for its own sake, (ii) feels one may justify costs to individual animal life or welfare in the expectation of future benefits to the conservation, management or welfare of that species or of some perceived benefit to humans, or (iii) feels the rights of the individual animal are paramount.

While evaluation of the *significance* of impact, or the relative balance of cost/benefit may be affected in this way by a personal ethical perspective, and costs in terms of psychological/emotional stress and suffering are rather harder to quantify, at least some of the explicit costs (as measured mortality, measured loss of individual fitness) may be objectively determined for many species for a number of these routine procedures of ecological fieldwork. In this paper I review some of that objective information to show the kinds of information that are available to the enquiring scientist: against which he/she may make a careful assessment of the precise welfare implications of each step of any such procedure – and thence make a reasoned decision on whether or not it may be justified. While the paper will address general issues, inevitably (and in some case to save embarrassment to others!) it will be biased towards my own experience.

Impact of simple behavioural observation

Disturbance to wildlife caused by simple intrusion into their environment (in walked census counts for example, or continuous behavioural observation), while apparently intervention at the lowest possible order, is not without impact. The effects of human disturbance (largely recreational disturbance, but the same principles apply) on wildlife have been explicitly examined by Freddy *et al.* (1986), Jeppesen (1987), Tyler (1991), van der Zande *et al.* (1984) – and see review by Putman and Langbein (1992). Both short-term responses to disturbance with no lasting effect and long term changes in breeding performance, for example, have been reported. Ornithologists have long acknowledged that human intrusion can influence social behaviour, reproductive performance of adults and survival of chicks (e.g. Duffy, 1979; Anderson and Keith, 1980; additional references reviewed by Farnsworth and Rosovsky, 1993).

Distress and mortality during capture and tagging operations

As noted earlier, it is generally extremely hard to determine what may constitute 'distress' in animals – and certainly difficult to quantify this rather nebulous concept of 'suffering' – a problem frequently recurring in debates over the welfare implications in housing, handling or management of farm animals, animals in zoos or circuses (Broom and Johnson, 1994). A number of behavioural protocols have been suggested however (e.g. Dawkins, 1980, 1990; Bateson, 1991; Wiepkema and Koolhaas, 1992) and some progress has been made in domestic mammals in the measurement of levels of cortisol production (Smith and Dobson, 1991) or blood levels of breakdown products such as LDH (lactic dehydrogenase)-5.

Initial attempts to apply these same physiological indicators in measurements of the perceived stress of capture and handling procedures for fallow deer (*Dama dama*) have been made by Jones and Price (1992). Results of these studies show that levels of LDH-5 are indeed elevated immediately after capture and decline as the animals become more and more quiescent after handling and prior to release. However, the LDH levels recorded are tremendously variable and it proves impossible at present to develop any predictive model of levels of stress implied by a given recorded blood concentration of LDH-5 (Jones and Price, 1992). Further difficulties in interpretation arise in that levels of this enzyme increase as the result of prolonged physical exertion, as the result of physical bruising, as well as in response to 'psychological' stress and thus it seems unlikely to develop as a reliable indicator against which to calibrate the stress levels imposed by different procedures.

Levels of actual physical injury or mortality contingent upon different procedures are more readily quantified. Continuing the cervid theme, I choose as my illustrations here, data on injuries and mortality rates associated with different methods of live capture of British deer. The statutory organization responsible for issuing licences for live capture of deer, the Nature Conservancy Council, and its daughter organizations (EN, CCW, SNH) all require a written return from licensees after each exercise but, few comments in these reports refer explicitly to injury and mortality rates. Data presented here derive from analysis of my own various catches of fallow deer in recent years. Only physical methods of capture were employed rather than chemical methods of immobilization (see for example reviews by Jones, 1984 and Harrington, 1991). Fallow deer were caught either in long-nets (Cockburn, 1976; Smith, 1980) or within purpose-built wooden handling units. Between 1985 and 1992 we caught and handled nearly 5500 animals during winter catching exercises; 670 of these were handled in fixed catchups, 4750 were caught by netting. Both methods result in some bruising and minor lacerations to a proportion of animals (not recorded).

Major injuries resulted in humane destruction of a total of 56 animals during the entire 7-year period (1.03%); serious injury or death was higher (1.33%) in purpose-built handling units than where animals were caught in long-nets (0.99%).

Equivalent data are available from the US: in review of levels of mortality or serious injury associated with different methods of physical capture of white-tailed deer (*Odocoileus virginianus*) Sullivan *et al.* (1991) recorded mortality rates of between 2.1% and 16.2% amongst deer caught by box trap or corral trap (all statistics here are restricted to studies with large samples > 100); mortalities of around 6–7% with the use of drop nets or cannon nets, and mortalities of around 0.9% associated with drive-netting techniques where deer are driven by beaters or helicopter into long-nets. Mortalities experienced in all these methods were considerably higher when deer were tranquillized post-capture (Sullivan *et al.*, 1991) and certainly mixing chemical methods of restraint with physical capture is almost always counterindicated (but see Chapman *et al.*, 1987 in relation to capture of muntjac deer).

One of the main problems here is accurate assessment of mortality arising from such capture operations *because not all consequential mortality is immediately apparent at the time of the catch-up*. Deer, like a number of other ungulates, hares, and many birds are subject to trauma-related stress-shock, most commonly manifested in the form of a progressive post-traumatic myopathy (Haarthorn *et al.*, 1976; Chalmers and Barrett, 1982; Putman, 1990) in which rapid and irreversible changes in muscle tissues lead to prostration, progressive depression and death after a period of up to 4 days after capture. Clearly, there are marked differences in mortality rate associated with different capture methods available, but my main point in this context is rather that *the data on which to draw such conclusions are readily available if we choose to seek them out.* Even the most favoured of the techniques (drive-netting), was accompanied by mortality rates of ~ 1% in both the studies summarised by Sullivan *et al.* (1991).

In the UK, licences are also required under the Wildlife and Countryside Act for use of live-capture traps to take small mammals if the traps are set in such a way that possibly they take shrews. Such provisions were enacted because, whatever efforts are made to provide adequate or appropriate food for shrews taken within such traps, mortality rates of animals trapped are very high. Even when used just for sampling more abundant species such as voles and mice, some incidental mortality is accepted as inevitable. Sadly little formal data are available on such mortality among small mammals during population studies by, for example, Longworth trapping; but my own experience suggests once again that while with care, it may be kept to low levels (Gurnell and Flowerdew, 1982), it is nonetheless significant and certainly not zero.

Small mammals frequently also perish by drowning in pitfall traps set out to sample ground active insects. Such observation, together with the incidental mortality of shrews in Longworth traps set for other species, already noted, make the additional point that evaluation of the impact of any procedure should extend to consideration of injury and mortality within non-target species. (These latter two procedures: live-trapping of small mammals and pitfall-trapping for invertebrates are methods commonly employed during educational field courses as well as in primary research, and emphasise again the point that even where no specific licence may be required permitting exercise of some restricted procedure for educational purposes, everyday sampling regimes used in teaching exercises may well raise the same ethical issues of justification as academic research. Here is an opportunity to explore these same issues while introducing the techniques themselves to students.)

Ethical issues associated with catching and ringing birds are considered by Greenwood (1994). Although the British Trust for Ornithology as the main body responsible for regulating such activities does not routinely gather mortality data, incidental mortality is considered infrequent because training standards required for UK ringers are extremely high and individual licences are not issued to ringers until they have served an extensive apprenticeship under the supervision of more experienced workers. Further, new methods are introduced cautiously through small numbers of very experienced ringers, exploring carefully the difficulties and dangers before they are adopted more widely and thus ensuring that potential problems are generally identified without mortality incidents.

Capture and handling of juveniles. One special case of capture for marking perhaps warrants separate consideration here: the case where mammalian neonates are marked soon after parturition or birds are ringed at the nest either as pulli or as adults. The relative immobility and defencelessness of neonates, coupled with close seasonal synchronisation of breeding facilitates the capture and marking of large numbers of individuals in a short space of time. For a number of bird species capture even of adults at the nest may be appropriate because the birds cannot conveniently be approached or capture at any other time. Clearly a potential problem here is maternal rejection of juveniles handled and thus scented by humans or nest desertion due to disturbance.

Drawing once again from my own experience, where we have routinely marked under licence large numbers of red deer calves and fallow fawns within hours of birth (although always restricting such marking until after the neonate has been licked clean of birth membranes and is fully dry), subsequent monitoring of mortality rates amongst marked and unmarked fawns showed no significant difference in mortality; similar conclusions are reported by Ozoga and Clute (1988) for white-tailed deer. The implications of catching and marking adult birds at the nest are reviewed by Kania (1992).

Loss of fitness caused by marking, tagging or sampling

Handling of any animals for sampling whether for simple measurements of body-weight, tibia or winglength, examination of reproductive status, or more intrusive sampling such as taking blood or faecal samples, involves a certain measure of stress. Many of the more intrusive procedures properly require special licensing, but all demand restraint and manipulation of wild animals to whom close proximity to humans is necessarily traumatic. In the extreme, prolonging handling for such procedures may cause sufficient trauma as to result in death directly, or as a result of later myopathic degeneration.

Marking of captured animals prior to release may also involve extended handling. Such marking may serve a variety of purposes: animals may be marked to distinguish animals previously captured from new capture in Capture-Mark-Recapture techniques; they may be marked to ensure future recognition of known individuals over the short or long term; or, they may be fitted with radio-telemetric devices for future location/triangulation. Requirements of the marking programme may determine the length of handling required to affix the tag; equally the mark itself may cause immediate or subsequent distress.

Most marking of vertebrates for simple distinction of previously captured individuals from new captures or to provide identification at the individual level in the short term, can be achieved by non-intrusive means: fur-clipping or dyeing for small (Gurnell and Flowerdew, 1982), or larger mammals, scale-clipping or paintmarking in reptiles or amphibians (Blanc and Carpenter, 1969; Spellerberg, 1977; Simon and Bissinger, 1983; reviewed by Swingland, 1978; Ferner, 1979). Such marks are lost however during any subsequent moult or skin change and will not persist from year to year.

Permanent marking for identification, for birds and larger terrestrial mammals, is most commonly by attachment of some form of visible tag (leg ring, wingtag, eartag or coloured collar); marking of smaller mammals was traditionally achieved through toe-clipping – in amputation of particular combinations of digits (Twigg, 1975 gives a full review; see also Fairley, 1982). Toe-clipping historically has also been the method most extensively used in permanent marking of reptiles and amphibians (Bellairs and Bryant, 1968). But use of such procedures is now generally regarded with disfavour and actively discouraged by professional societies. (Toe clipping is, in any case, not an infallible technique from the researcher's viewpoint; toes may be lost naturally, confusing subsequent identification; in amphibians, amputated digits regenerate over a period of as little as 7 months: Ferner, 1979).

Alternatives to toe-clipping which have been exploited include eartagging, leg-ringing or, for medium sized mammals such as rabbits or squirrels as well as for larger reptiles and amphibians, hot- or cold-branding (e.g. Daugherty, 1976). More recently some experimental work has been done to assess the potential for field use of subcutaneous electronic tags based on transponders. Such tags have been used successfully for individual mice and rats in the laboratory, although not yet proven for field use. Certainly subcutaneous tags seem the only practicable way of marking animals with no obvious external appendages or with highly streamlined bodies (such as snakes, or aquatic mammals) or others living in dense habitats who might otherwise regularly snag neck collars or eartags.

While some instantaneous injury is inflicted in hot or cold branding, insertion of external eartags or in the surgery involved in implantation of internal tags, the trauma seems shortlived. More problematical seems subsequent injury or loss of efficiency through wearing such markers. Frequent injuries used to be reported from ill-fitting leg rings among birds and more particularly damage caused in wading or water birds during winter from ice accumulation. Williams *et al.* (1993) provide a recent analysis of the effects of mass-ringing on body condition and gosling survival of Lesser Snow geese, while Calvo and Furness (1992) consider the effects of different marking devices.

Southern (1964) considers in detail the difficulties associated with using leg-rings in marking studies of small wild rodents and it is clear that here too ill-fitting rings can lead to injury, with swelling of the limb around the ring and even subsequent loss of the foot or leg. Fullagar and Jewell (1965) also offer a comparison of the effectiveness of a range of different tagging methods for small mammals and highlight problems associated with the use of such leg rings (though neither study gives precise figures). In a study on the survival of individual woodmice and bank voles marked by toe-clipping or small metal ear tags – and the survival of the mark itself – Hill (1985) noted that there was no significant difference in survival (survival rate or actual longevity) of animals marked by eartag against those marked traditionally by toe-clipping, that eartags were retained by 91% of all animals marked and that in those cases where tags were subsequently lost, the average time to loss was 1.6 months in *Apodemus sylvaticus* and 3.0 months in *Clethrionomys glareolus*.

Although tag-survival rate was high in this particular case, Hill's study draws explicit attention to the compromise that must be faced by any field worker between using the marking system that will minimize any risk to the animal and yet fitting a tag or marker that will persist for a sufficient period to permit ready identification.

One additional problem associated specifically with the fixing of radiotelemetric transmitters to study an animal is caused by the necessary weight of such devices and the increased load carried by marked animals which can further seriously restrict movement or impede locomotion, reducing fitness, possibly increasing probabilities of mortality by reducing efficiency of foraging or hindering escape from predators. In radiotracking studies it is necessary to give thought to the size of the transmitter package and the period over which the animal will have to carry it (which may well be its natural lifespan if recapture is unlikely). As a general rule of thumb it is argued that a transmitter package should weigh no more more than 3% of the animal's own unladen bodyweight. A number of authors have attempted to investigate the effect of affixing radio collars to their study animals in this way by investigating body weight changes or changes in behaviour. Excellent reviews have been offered by Kenward (1987) and White and Garrott (1990).

One final issue rarely considered explicitly by those involved in marking studies of any kind is the influence of capture and handling, or any mark applied, on other conspecifics and thus the released individual's social relationships. Human scent associated with handling and any transient or longer-term changes in behaviour due to stress of handling, to the after affects of drugs, or simply associated with the wearing of a collar or tag may well result in a wary reception amongst conspecifics, or possible rejection. The mark itself may directly affect social acceptance by conspecifics. In animals with more formally-structured social organization, such marks may affect dominance rank; and for both solitary and social species may possibly influence future reproductive fitness – in both cases conferring advantage through the appeal of novelty, or disadvantage through suspicion of the unfamiliar.

I have no evidence for social advantage or disadvantage resulting from marking. As noted earlier, we have had no cases of rejection of deer fawns handled by us and marked as neonates, though it should be stressed that handling was always kept to a minimum and no animals were ever marked before being licked dry by the mother. Among the social species with which I have worked we have never seen any suspicion by other conspecifics of adult fallow marked by us with eartags or collars. However, the major problem experienced by one researcher working with Reeves' muntjac is that the lifespan of the eartags is limited due to them being chewed by other muntjac until they are lost or unreadable (S. Harris, personal communication). I have found similar problems in populations of park fallow deer. Such observations certainly suggest the tag is noticed and is an object of special attention, but are hardly consistent with social rejection.

Conclusion

This paper has inevitably covered a wide range of issues. My purpose throughout has been to enhance awareness of the issues involved in handling and marking wild animals during routine ecological field studies: to make explicit many of the implications which many may not appreciate, or at best overlook. While we may not yet formally assess the extent of distress experienced by animals during such procedures, I hope to draw attention to the fact that objective data are frequently available on the risk of actual physical injury/mortality associated with any particular operation and the relative merits and demerits of alternative methodologies so that the responsible fieldworker may formally assess the risks involved in adopting a given procedure. Equally, it is clear that, for many of the procedures, while choice of the best available alternative may *minimize* risk of mortality/injury, *all* available methods are associated with at least some level of risk greater than zero. Therefore there remains always the subsidiary question of what is an 'acceptable' level of 'suffering', of mortality or fitness loss beyond which one should abandon the research?

Any framework for individual decision-making must mimic the cost-benefit analysis which is carried out in assessment of any procedures formally requiring statutory licence, but I believe the same assessment of risks and benefits should be applied to any piece of work whether requiring licence or not. In practice relatively few researchers explicitly confess to doing so (but see for example, Huntingford, 1984; Cuthill, 1990; and discussions of Dawkins and Gosling, 1992). As noted earlier, more formal approaches to cost-benefit analyses have been presented (for example, Smith and Boyd, 1991) but evaluation of even the statistics which may be collated about the risks associated with any given procedure and the benefits which may be gained, depends on one's individual moral viewpoint (as for example 'scientistic', 'utilitarian' or 'moralistic').

To focus the issues, therefore, I present a review of the decision stages *I* pass through in reaching ethical decisions in my own work, where I regularly capture, measure and mark deer. While I must satisfy myself on each point before I even embark on the whole programme of research, I rehearse them over and over again – even in the car *every* time I drive to each new catch-up; I believe one should never stop asking them.

(i) What are the objectives of the research programme on which I am embarking/in which I am engaged?
(ii) Can those objectives be realised in any other way other than by (in turn) (a) catching and measuring; (b) marking; (c) (if appropriate) blood sampling or other interventive procedure?
(iii) What alternative methods are available for capture, marking, restraint during sampling and sampling?

(iv) What are the actual risks of injury/death/discomfort associated with the different methods available for catching, marking, restraint and sampling for my particular species? And am I sure I have selected those methods which minimise such risks/costs?

(v) Given that some residual risk attaches to each procedure, does the value of the data I will collect during (a) catching and measuring, (b) marking, (c) sampling justify the possible cost in terms of injury or death?

(vi) If I am honest with myself: is my research simply self-indulgent, or do I honestly believe that the work that I am doing and the data that I will collect will genuinely enhance our understanding in an important area of biology, or, in my case, do I believe that the work will improve management for the future and thus are the risks I am taking with current animals' lives justified against the returns to be gained in the welfare of others in the future?

If I cannot satisfy myself on these points, I turn the car round.

Acknowledgements

I would like to express my thanks to Bob Carling for inviting me to put these ideas on paper for presentation to the INTECOL meeting in August 1994 – and then in the event, for reading it for me in my own absence. I am also most grateful to him for thoughtful comments on the manuscript itself. I would also thank Dr Arthur Lindley, Dr Jeremy Greenwood, Steve Whitbread and Nick Smith who also read various drafts or offered a number of very helpful suggestions.

References

Anderson, D.W. and Keith, J.O. (1980) The human influence on seabird nesting success: conservation implications. *Biol. Conserv.* **18**, 65–80.

Association for the Study of Animal Behaviour (1981) Guidelines for the use of animals in research. *Animal Behaviour* **29**, 1–2.

Bateson, P. (1991) Assessment of pain in animals. *Animal Behaviour* **42**, 827–39.

Bellairs, A. d'A. and Bryant, S.V. (1968) Effects of amputation of limbs and digits of lacertid lizards. *Anat. Rec.* **161**, 489–96.

Bender, E.A., Case, T.J. and Gilpin, M.E. (1984) Perturbation experiments in community ecology; theory and practice. *Ecology* **65**, 1–13.

Blanc, C.P. and Carpenter, C.C. (1969) Studies on the Iguanidae of Madagascar. III. Social and reproductive behaviour of *Chalarodon madagascariensis. J. Herpetol.* **3**, 125–34.

Broom, D.M. and Johnson, K.G. (1994) *Stress and Animal Welfare*. London: Chapman & Hall.

Calvo, B. and Furness, R.W. (1992) A review of the use and the effects of marks and devices on birds. *Ringing and Migration* **13**, 129–51.

Chalmers, G.A. and Barrett, M.W. (1982) Capture myopathy. In *Non-infectious Diseases in Wildlife* (G.L. Hoff and J.W. Davies, eds.) pp. 84–94. Ames, IA: Iowa State University Press.

Chapman, N.G., Claydon, K., Claydon, M. and Harris, S. (1987) Techniques for the safe and humane capture of free-living muntjac deer (*Muntiacus reevesi*). *Brit. Vet. J.* **143**, 35–43.

Cockburn, R.H.A. (1976) Catching roe deer alive in long-nets. *Deer* **3**, 434–40.

Cooper, M.E. (1987) *An Introduction to Animal Law*. London: Academic Press.

Cooper, M.E. (1991) British Mammals and the law. In *The Handbook of British Mammals, 3rd edn* (G.B. Corbet and S. Harris, eds.) pp. 24–34. Oxford: Blackwell.

Crowell, K.L. and Pimm, S.L. (1976) Competition and niche shifts of mice introduced onto small islands. *Oikos* **27**, 251–8.

Cuthill, I.C. (1990) Field experiments in animal behaviour: methods and ethics. *Animal Behaviour* **42**, 1007–14.

Daugherty, C.H. (1976) Freeze-branding as a technique for marking anurans. *Copeia* **4**, 836–8.

Dawkins, M.S. (1980) *Animal Suffering: the Science of Animal Welfare*. London: Chapman & Hall.

Dawkins, M.S. (1990) From an animal's point of view: motivation, fitness and animal welfare. *Behav. Brain Sci.* **13**, 1–61.

Dawkins, M.S. and Gosling, M. (1992) *Ethics in Research on Animal Behaviour.* London: Association for the Study of Animal Behaviour/Academic Press.

Diamond, J.M. (1983) Laboratory, field and natural experiments. *Nature* **304**, 586–7.

Diamond, J.M., Pimm, S.L., Gilpin, M.E. and LeCroy, M. (1989) Rapid evolution of character displacement in myzomelid honeyeaters. *Am. Nat.* **134**, 675–708.

Duffy, D.C. (1979) Human disturbance and breeding birds. *Auk* **96**, 815–16.

Fairley, J.S. (1982) Short term effects of ringing and toe-clipping on recaptures of woodmice. *J. Zool.* **197**, 295–7.

Farnsworth, E.J. and Rosovsky, J. (1993) The ethics of ecological field experimentation. *Conserv. Biol.* **7**, 463–72.

Ferner, J.W. (1979) *A review of marking techniques for amphibians and reptiles.* Society for the Study of Amphibians and Reptiles; Herpetological Circular **9**, 1–41.

Freddy, D.J., Bronaugh, W.M. and Fowler, M.C. (1986) Responses of mule deer to persons afoot and snowmobiles. *Wildlife Society Bulletin* **14**, 63–8.

Fullagar, P.J. and Jewell, P.A. (1965) Marking small rodents and the difficulties of using leg rings. *J. Zool.* **147**, 224–8.

Greenwood, J.J.D. (1994) Research on wild birds: ethical issues of ringing. Proceedings of the International Ornithological Congress, Vienna.

Gurnell, J. and Flowerdew, J.R. (1982) *Live Trapping Small Mammals – a Practical Guide.* London: Occasional Publications of the Mammal Society.

Haarthorn, A.M., van der Walt, K. and Young, E. (1976) Possible therapy for capture myopathy in captured wild animals. *Nature* **247**, 577.

Hairston, N.G. (1980) The experimental test of an analysis of field distributions: competition in terrestrial salamanders. *Ecology* **61**, 817–26.

Harrington, R. (1991) Guidelines for the capture and handling of deer. In *Methods For the Study of Large Mammals in Forest Ecosystems* (G.W.T.A. Groot Bruinderink and S.E. van Wieren, eds.) pp. 74–87. Arnhem, The Netherlands: Rijksinstituut voor Natuurbeheer.

Hill, S.D. (1985) *Influences of large herbivores on small rodents in the New Forest, Hampshire.* PhD thesis University of Southampton.

Huntingford, F.A. (1984) Some ethical issues raised by studies of predation and aggression. *Animal Behaviour* **32**, 210–15.

Jeppesen, J.L. (1987) The disturbing effects of orienteering and hunting on roe deer (*Capreolus capreolus*). *Danish Rev. Game Biol.* **13**, 1–24.

Jones, D.M. (1984) The capture and handling of deer. In *The Capture and Handling of Deer* (A.J.B. Rudge, ed.) pp. 34–85. Peterborough: Nature Conservancy Council.

Jones, A.R. and Price, S.E. (1992) Measuring the responses of fallow deer to disturbance. In *Biology of deer* (R.D. Brown, ed.) pp. 211–16. New York: Springer-Verlag.

Kania, W. (1992) Safety of catching adult European birds at the nest. *The Ring* **14**, 5–50.

Kellert, S.R. (1991) Japanese perceptions of wildlife. *Conserv. Biol.* **5**, 297–308.

Kenward, R.E. (1987) *Wildlife Radio-Tagging: Equipment, Field Techniques and Data Analysis.* London, New York: Academic Press.

Moulton, M.P. and Pimm, S.L. (1983) The introduced Hawaiian avifauna: biogeographic evidence for competition. *Am. Nat.* **121**, 669–90.

Moulton, M.P. and Pimm, S.L. (1986) The extent of competition in shaping an introduced avifauna. In *Community Ecology* (J.M. Diamond and T.J. Case, eds) pp. 80–97. New York: Harper and Row.

Ozoga, J.J. and Clute, R.K. (1988) Mortality rates of marked and unmarked fawns. *J. Wildlife Manag.* **52**, 549–51.

Paine, R.T. (1966) Food web complexity and species diversity. *Am. Nat.* **100**, 65–75.

Paine, R.T. (1969) A note on trophic complexity and community stability. *Am. Nat.* **103**, 91–3.

Putman, R.J. (1990) The care and rehabilitation of injured wild deer. *Deer* **8**, 31–5.

Putman, R.J. and Langbein, J. (1992) Behavioural responses of park red and fallow deer to disturbance and effects on population performance. *Animal Welfare* **1**, 19–38.

Simon, C.A. and Bessinger, B.E. (1983) Paint marking lizards: does it affect survival? *J. Herpetol.* **17**, 184–6.

Smith R.H. (1980) The capture of deer for radio-tagging. In *A Handbook on Biotelemetry and Radiotracking* (C.J. Amlaner and D.W. Macdonald, eds) pp. 313–17. Oxford: Pergamon Press.

Smith, J. and Boyd, K. (1991) *Lives in the Balance: the Ethics of Using Animals in Biomedical Research.* Oxford: Oxford University Press.

Smith, R.F. and Dobson, H. (1990) Effects of preslaughter experience on behaviour, plasma cortisol and muscle pH in farmed red deer. *Vet. Rec.* **126**, 155–8.

Southern H.N., ed. (1964) *The Handbook of British Mammals, 1st edn.* Oxford: Blackwell.

Spellerberg, I.F. (1977) Marking live snakes for identification of individuals in population studies. *J. App. Ecol.* **14**, 137–8.

Strauss, S.Y. (1991) Indirect effects in community ecology: their definition, study and importance. *Trends Ecol. Evol.* **6**, 206–10.

Strong, D.R., Simberloff, D., Abele, L.G. and Thistle A.B. (1984) *Ecological Communities: Conceptual Issues and the Evidence.* Princeton, NJ: Princeton University Press.

Sullivan, J.B., De Young, C.A., Beasom, S.L., Heffelfinger, J.R., Coughlin, S.P. and Hellickson, M.W. (1991) Drive-netting deer – incidence of mortality. *Wildlife Society Bulletin* **19**, 393–6.

Swingland, I.R. (1978) Marking reptiles. In *Animal Marking: Recognition Marking of Animals in Research* (B. Stonehouse, ed.) pp. 119–41. London: Academic Press.

Twigg, G.I. (1975) Marking mammals. *Mammal Rev.* **5**, 101–16.

Tyler, N.J.C. (1991) Short-term responses of Svalbard reindeer to direct provocation by a snowmobile. *Biol. Conserv.* **56**, 179–94.

van der Zande, A.N., Berkhuizen, J.C., van Latesteijn, H.C., ter Keurs, W.J. and Poppelaars, A.J. (1984) Impact of outdoor recreation on the density of a number of breeding bird species in woods adjacent to urban residential areas. *Biol. Conserv.* **30**, 1–39.

White, G.C. and Garrott, R.A. (1990) *Analysis of Wildlife Tracking Data.* London, New York: Academic Press.

Wiens, J.A. (1989) *The Ecology of Bird Communities, Volume 2: Processes and Variations.* Cambridge: Cambridge University Press.

Wiepkema, P.R. and Koolhaas, J.M. (1992) Stress and animal welfare. *Animal Welfare* **2**, 195–218.

Williams, T.D., Cooke, F., Cooch, E.G. and Rockwell, R.F. (1993). Body condition and gosling survival in mass-banded Lesser Snow geese. *J. Wildlife Manag.* **57**, 555–62.

12

Wildlife conservation in churchyards: a case-study in ethical judgements

NIGEL S. COOPER
The Rectory, 40 Church Road, Rivenhall, Witham, Essex, CM8 3PQ, UK

Groups promoting wildlife in churchyards, or other sites, discover that they face normative questions that have no scientific answers. The language of management is used for handling these questions, but this metaphor has unhelpful associations with predetermined goals, a culture of control and self-centredness. Using a case-study approach, conflicts between conserving natural entities or natural processes (e.g. transplanting scarce plants); between caring for the individual organism or for the system (e.g. felling trees); and between conserving the natural or the cultural heritage (e.g. repointing walls) are examined. These cases of conflicts of duty illustrate the value of attention to circumstances, proportionality, and compromise. The social mechanisms of moral debate include legal protection and the power of stories to give meaning and vision. Ethics is a communal activity. By listening to others and attending to nature our sensibilities will become more refined and our ethical judgements will develop.

Keywords: biodiversity; management; conflicts; casuistry; communal ethics

The importance of churchards for biodiversity

When you walk into a typical English churchyard, you are greeted by a sense of the great diversity of life there. [Churchyards are the lands around church buildings where, often, people are buried, but I am using the word in an extended sense to include land around any places of worship and also any land used for human burials. Although I have been concerned with this wide range of sites, my principal experience has been with the Church of England. There are particular features to the conservation of wildlife in churchyards, but many of the principles apply to other sites where conservation has to be married to other uses, e.g. farmland, and even to sites that are primarily nature reserves.] Typically, there is a wide range of trees. The subtle texture of the 'grass' is created by the variety of leaf-forms that make up the sward. The stonework is decorated by a pastel pattern of lichens and mosses. This sensation of biodiversity is confirmed by studies, largely unpublished, of the importance of churchyard sites for rare and uncommon species. [Examples include records of flowering plants by Bob Leaney for the Norfolk Wildlife Trust, 72 Cathedral Close, Norwich; records of Tom Chester of the British Lichen Society, Natural History Museum, London; and records of basidiomycetes by Tony Boniface of the Essex Field Club, and of bryophytes by Tim Pyner, both c/o the author. The Living Churchyard Project (see below) has contacts with other recorders for churchyards including David Glue for birds, Susan Clarke for butterflies and Pat Donovan for plants in Sussex.] This biodiversity arises because churchyards offer wildlife interesting habitats in exposed stone and old trees. Most importantly, their grassland has not usually been improved by fertilizers, herbicides, or ploughing and reseeding. When so much of Britain's ancient grassland has been lost, churchyards are an important remnant (Dennis, 1993). Burial grounds and religious sites elsewhere in the world are also of value to conservation. [For details of examples contact The Alliance of Religions and Conservation, c/o ICOREC, Manchester Metropolitan University, 799 Wilmslow Road, Manchester M20 2RR, UK.]

For the Church of England, which is responsible for the majority of churchyards of wildlife interest in England, there has been a growing awareness of their importance. The first publication was by Barker (1972). The third edition of *The Churchyards Handbook* (Burman and Stapleton, 1988) contained much general advice on wildlife for a wide audience. And the Council for the Care of Churches has now commissioned another booklet on wildlife care (Cooper, 1995). Ecumenically, the Living Churchyard Project was launched in 1988 and this has organised and stimulated much churchyard conservation work across the churches. The Living Churchyard Project produces literature and advice. It is based at the Arthur Rank Centre, National Agricultural Centre, Stoneleigh, Warwickshire CV8 2LZ, UK. Many counties, or dioceses (the Church of England equivalents), now have their own groups to promote wildlife conservation in churchyards. Local authorities in charge of burial grounds are also involved.

In addition to providing habitats for plants and animals, churchyards are also significant for the conservation movement because of their public profile. Promoting churchyard conservation engages in debate a wide range of people responsible for these sites: those who cut the grass, those who sit on parochial church councils who are responsible in law for the maintenance of churchyards, and diocesan officers. Beyond these people, who are actively involved, are the millions of people who visit churchyards to tend graves or to come to church, as well as those who pass by, looking over the churchyard wall or hedge. Even visiting churchyards just for their atmosphere and to read old gravestones is a popular pastime. If these people see a good example of nature conservation in harmony with human use, this will help to persuade them that this is possible and desirable. Arguably, churchyards are amongst the most visited sites where nature conservation may be practised.

Churchyards also carry great symbolic weight because of their religious significance and their use for human burial. How people treat them may be an expression of their ideals, perhaps how they believe that God would wish the world to be if it were not damaged by sin and evil. The diversity of churchyard styles witnesses to the very different utopia people believe in. Where beauty and hope are identified with a masterful control and a military tidiness, the churchyards are hardly distinguishable from old-fashioned municipal gardens. Where faith in the future has gone, as in some depopulated rural areas, all efforts to intervene are abandoned and the monuments fade from recollection as they become overgrown. Conservationists have the opportunity in churchyards to pass on their alternative vision of humans allowing wildlife space to flourish in places primarily used to worship God and remember the dead. Yet even among conservationists there are competing ideals. Some favour keeping the grass short to promote plant species diversity: others wish to leave the grass to grow long to benefit the animals, especially insects. This is just one difference amongst those who share the same basic commitment to care for wildlife.

Moral questions raised

That there are these alternative visions alerts us to the role of values in conservation. Those who care for churchyards, and those who advise them, face questions such as, 'What should we do or not do here and now?'; and the word 'should' indicates that a normative judgement is being called for. Similar questions face those conserving wildlife in other types of sites. These questions in turn raise the more general question: 'How should we make ethical judgements in nature conservation?'

Sometimes it is assumed that science can provide the answer to what should be done. But can science answer our general question about how to make ethical judgments? Science offers answers to many crucial questions in conservation, but not to normative ones. It describes what is there and how rare or threatened it is. It predicts what activities will promote or harm a population of a species. It suggests methods for changing a habitat in a chosen direction, or for keeping it relatively unchanged. But science alone cannot determine what ought to be done. Doubt may be raised about how secure a distinction can be made between facts and values, 'is' and 'ought'. But certainly statements of scientific fact do not straightforwardly generate statements of value. Yet there is still a tendency among conservationists to feel that science does tell us what to do. Ethics then becomes a tool to persuade other non-scientific persons to adopt the behaviour scientists are advocating (Grove-White and Szerszynski, 1992). As an example of this sort of scientism, Hambler and Speight (1995) argue that traditional management, such as coppicing, ought not to be reintroduced on scientific grounds:

> (E)cology and the precision of expression it encourages are beginning to force logically compelling, if unpalatable, conclusions onto managers and the public (p. 137).

If there are agreed aims, e.g. increasing the number of invertebrate species, then science may direct that restoring traditional management is counter-productive. Hambler and Speight ignore, however, the diverse aims that are held by people involved in conservation. Some will agree that invertebrate biodiversity should be maximized, others will argue for restoring historic landscapes, others may be concerned to enhance beauty through increasing the show of butterflies and spring flowers. Hambler and Speight concede, 'Confusingly, different scientists may recommend different managements'. Perhaps it is not only aims that differ, but that science itself is uncertain.

How does this assumption that science gives the answers come about? Grove-White and Szerszynski (1992) are among those who claim that values determine facts, or at least that there is no natural description of the world prior to normative commitments (Howarth, 1995, in this publication discusses this position in greater detail). This may be hard for scientists to accept, and it does deny the fact/value dichotomy from the other direction.

Scientists may more readily agree that alongisde their science there is a set of value judgements that are often hidden from view. Such a set might include an attachment to the study objects, a passionate love of some aspect of the wild that led them to take up the relevant branch of science. It might include the feeling that what is rare is interesting and so worth saving. There may be an assumption that the human is separate from the natural world, either, as in agricultural science, making human needs and desires paramount, or, as in traditional ecology, seeking out for study natural systems where human interference in the natural process is minimized. Sometimes the value judgement lies unnoticed in the unexpressed protasis; before each 'should' and 'ought' there may lurk an invisible conditional clause: (if you value X), you must do Y. Another way value assumptions are often hidden from us is in the choice of metaphors used in a discourse (Ruse, 1994). The metaphor of management is commonly used in practical wildlife conservation. What are the values implicit in this metaphor?

The metaphor of management

Environmental management, management plans, site managers – again and again the word management is used in conservation. It is believed that we cannot avoid involvement with the natural world, but we should at least manage it well and not negligently. The recommended steps in site management are: first survey to discover what is there, second draw up a management plan and act upon it, and third record results and review the plan in the light of them. This is a good process as it incorporates the epistemic values of logical thought, consistency, and effective decision making. But the methods of management hide the non-epistemic, non-scientific judgements involved in managing a site.

And this is not surprising, as the metaphor of management comes from business, not the source of first choice for a method of moral debate. I am not suggesting that the business world is immoral, but that, like the scientific world, it is a world where values are hidden from view, and the language of management may even mislead conservationists, because it carries several implicit values which may not suit nature conservation. Three such values particularly cause concern:

(i) Management assumes that the purpose of the enterprise has already been determined. Managers work to targets set by others. There is the simple assumption that the business is to make profits. This may be appropriate for business; it is not appropriate for nature conservation. It leaves open the most crucial question, 'What is the goal the management is expected to achieve?' Of course, most reserve management plans include a section on aims, but the point is that the model of management itself provides no help on how to set these aims. What might be the aims for a churchyard? Is it to be primarily a fitting place for burial, or to make a good backdrop for the church, or to provide sanctuary for a scarce organism, or to mimic a mediaeval meadow, or to maximise species diversity, or to be part of a network of sites available for meta-populations of a species to disperse among, or for natural processes like wood decay to operate unhindered, or ...? The list is almost endless. Whichever aim is set, a good manager of a churchyard could come close to achieving it, but, as a manager *per se*, she or he could not say which aim should be chosen.

But more disturbingly still, the management model tends to eclipse the discussion of ends. MacIntyre (1981) argues that the modern concept of management presupposes the belief that there can be no rational argument about values and ends, that value statements are no more than the emotive verbal expression of preferences and feelings. If it is believed that questions of aims are questions of values, and on values reason is silent, all the debate becomes concentrated on what reason can discuss, that is, the effectiveness of management. This in turn means that competing views come to be compared on the basis of efficiency and not on their moral worth. Those who can demonstrate effective control and the successful operation of power are thereby considered worthy to be given control over management, and this without any reference to their aims.

(ii) Management, then, assumes a culture of control. The manager sees to it that decisions made are implemented and the set targets are achieved. The workforce may be invited to participate in their own management, but only if they accept working within the overall aims set for them. But the conservationist is in a dilemma. The more humans attempt to control the world, the less natural it becomes. On some understandings, control is antithetical to preservation. On all understandings, the natural world refuses to accept control and events continually surprise us. Weather, population sizes, migrations etc. all prove a refractory workforce.

In churchyards there is an additional factor. Places of burial raise the issue of control in an acute form. Death is the ultimate assertion in nature that we are not in control. This can provoke a reaction on the part of churchyard managers who then allow nothing to grow there without permission, and even then it must grow tidily. Those who wish to let nature be can appeal to a religious hope, that new life comes from death: the aptly named life-cycle. A certain wildness in a burial ground expresses this belief in the regeneration of life (Cooper, 1989).

(iii) Management is a divisive and self-centred notion. The management team is differentiated from the rest of the workforce. The management of a company is conducted in competition with other companies. Conservation, as a social force, needs to be a co-operative and community venture if it is to succeed, involving local people. And biologically, rather than be seen in isolation, the managed sites are more accurately considered as part of a landscape with a myriad of interactions with sites near and far away (both these points are emphasized by Idle, 1995).

This criticism may seem pathetic in comparison with the dominant position of management terminology in conservation. My hope may not be so much to supplant management as a metaphor, as at least to show that more is needed. Management is inadequate because it hides moral assumptions, those assumptions may not fit easily with conservation, and the management model does not accurately describe what really goes on in conservation. [A somewhat similar criticism could be made of that other popular term, stewardship. Stewards, like managers, work to values determined elsewhere which they then accept; stewards are part of a hierarchy of control, where the non-human is directly dominated but the stewards themselves are only accountable in the distant future; stewardship assumes a metaphysical division of humans from the rest of nature (Palmer, 1992).]

The case-study approach

As an alternative, I offer the method of case-studies. [This approach is in sympathy with whose who believe an impatience with natural history, which is also based on case-studies, is inappropriate for ecologists (e.g. Schrader-Frechette and McCoy, 1993; Weiner, 1995), as opposed to those who wish to make ecology a hard, predictive science (e.g. Peters, 1991).] Used as an investigative method, it should illuminate how moral decisions are made in conservation, and what principles underlie them. What is revealed is a case-structured approach to moral reasoning, i.e. casuistry.

> Nowadays the moral problems of public policy are not merely stated in casuistical ways: they are also debated in the same taxonomic terms, and resolved by the same methods of paradigm and analogy that are familiar to students of common law and casuistry (Jonsen and Toulmin, 1988, p. 306).

The metaphor of cases is a legal one, and that helps to make ethical judgements explicit. I have used ideas from the religious parallel of casuistry. This is a method for giving moral advice to people in the particular predicaments in which they find themselves (Kirk, 1936; Dewar, 1968; ACCM, 1974). [Casuistry as a method takes no sides in debates such as whether morality is subjective or objective, or whether there is a single ethical principle as the fount of the whole moral system (monism) or whether there is a plurality of principles and moral systems. Environmental philosophers are particularly concerned about this last issue. Pluralism is attractive because it opens up the possibility of new principles to cherish sentient animals and the rest of nature (e.g. Stone, 1988). It is feared because it may generate contradictory rulings allowing self-interest to sway the resolution of these conflicts (Callicott, 1990).] Historically, since Blaise Pascal, it has been unfairly criticised for hypocrisy and sophistry. It also has some genuine limitations, being rather static, individualistic, and legalistic. But it does offer a model for applying ethics to actual situations. Rather than taking a deductive, top-down approach to ethics, working everything out from first principles (and often failing to get as far down as the moral decisions we have to make in ordinary life), casuistry begins with actual cases and tries to trace the lines of argument from cases to principles in both directions. Paradigms would be a better word here than principles. The casuist accepts a series of paradigmatic cases that instantiate moral principles and that receive broad assent. Each case for consideration is then related to this taxonomic series of precedents by analogy and the application of second-order maxims (Jonsen and Toulmin, 1988).

Sample cases

Whether to introduce new plants

In Essex, UK is a local group that advises on churchyard conservation. In this group we have debated whether it is right to introduce new plants into a churchyard, e.g., to take a clump of Green-winged Orchid, *Orchis morio*, from a turf that was about to be destroyed and to translocate it to a nearby churchyard; or would it be right to plant a Wild Service Tree, *Sorbus torminalis*, in a gap in a hawthorn hedge, *Crataegus monogyna*; or to use commercial wild-flower seed to enrich the species composition of a lawn, or to decorate a grave-spoil heap.

The arguments which have been used have been diverse. In favour of introductions it can be said that they give some particular plants a chance of survival, or aid the dispersal of a scarce species (particularly with climate change in prospect), or provide a diversified habitat for other organisms and make a good show for the public, particularly in urban areas. Against, it is argued that moving the plants is merely palliative and does not help the socially invisible species (e.g. in soil and litter) that will be destroyed by a developer, or that introductions will obscure natural distribution patterns with a history of their own, or they will damage at least the integrity of the community that was already there even if they do not get out of control or bring new diseases with them (other arguments also could be used, e.g. Maunder, 1992).

As these cases have been debated, it has become clear that we are facing a conflict between two principles. This is an example of what in casuistry is termed a perplexity, a case of a conflict of (*prima facie*) duties. Many ecologists wish to conserve species *and* to allow natural processes to operate without interference. But these natural processes may lead to the extinction of a species, at least locally. So the perplexity arises, should one conserve a particular species by interfering in natural processes or allow the processes free rein and risk the loss of the species? The conflict of duties to maintain natural entities, anything from individual organisms to ecosystems, and to maintain natural processes, including succession, death and extinction, is a lively topic among philosophers too as is illustrated by Holland (1995) and Comstock (1995).

Our discussion has helped to clarify when it is appropriate to help a species or a population and when to leave it to its own devices. We have each come with initial preferences. Amateur naturalists have tended to defend the plants themselves as representatives of their species, the professional ecologists have been more concerned with the evidence for processes, such as distribution patterns. These initial positions have become modified as we have discovered that the answer usually depends on the site in question. To resolve the perplexity we often use a maxim like 'build on strengths'. Where a site shows evidence of continuity in its plant community, with indicator species of ancient grassland, it would be wrong to disturb the continuity by introducing new species. On the other hand, in an urban amenity grassland, it would be valuable to diversify it by reconstructing an approximation of ancient grassland which could help educate many people.

However, the maxim 'build on strengths' is too simple. It does not take sufficient account of what lies around the site. What is required is more akin to landscape planning, looking at each site not in isolation but in its context at different scales. For example, one planning consideration is the balance of different types of conservation sites in an area. Botkin (1990), writing in a North American context, gives examples of three different types of conservation sites: no-action wilderness, preagricultural wilderness, and conservation areas for specific features of biodiversity. In Britain the human impact on the environment has been intense for a long while (Rackham, 1986). But there can still be different types of nature reserve; priority can be given to natural processes, or to a specified habitat or species, or to a historic landscape or traditional management. Hence, in a particular churchyard priority might be given to maintaining a population of a rare plant. More often, priority in a churchyard will be given to the grass sward as representative of grassland landscape and habitat in continuity with that before the agricultural revolution. What other sites exist nearby will be one of the factors in deciding the conservation priorities of a churchyard.

Another consideration is how churchyards might assist the dispersal of species. What can they contribute to a local patchwork of sites? In this case the scale of the organisms selected for concern, and the scale of their dispersal strategies, will be crucial. The practical use and spiritual and emotional investment in a particular churchyard will also affect the choice of priorities.

Certainly we have learned that the old maxim is true, that circumstances make the case, that the setting makes the site.

Whether to fell a tree

Trees are special organisms for many people. Our conservation group shares this widespread conviction, but we are also aware that it is the remnants of ancient grassland that make churchyards special for wildlife and these are threatened by the shade of expanding tree cover. Sometimes we even discuss whether a tree should be felled for the sake of the grassland. This debate is akin to the lively debate over whether the culling of animals is ever justified to protect a habitat (e.g. Midgley, 1992; Ehrenfeld, 1993). In other words, what may protect the individual against the demands of the system? If there are some accepted principles that protect individual humans, can these be extended to cover animals, perhaps on the grounds of their sentience? Is it in any way possible to extend them to trees, on the grounds of their longevity, their majestic size, their aesthetic appeal (cf. Stone, 1974)? If they can be extended to some degree, in what ways are these principles modified? Casuistically, these are questions of moral doubt, 'Does the principle apply in this extended case?' 'Can the analogy be legitimately extended?'

Society offers some answers to these questions in the form of English secular and ecclesiastical law. Individual trees may be protected by Tree Preservation Orders so that they may be felled only with the permission of the local authority. Since the Care of Churches and Faculty Jurisdiction Measure 1991, diocesan chancellors (the principals of diocesan consistory courts) are required to produce guidance on how permission to fell trees in churchyards is to be obtained. In my diocese of Chelmsford, Essex, complex guidelines were issued in 1994. (These are obtainable from the Chelmsford Diocesan Advisory Committee, Guy Harlings, New Street, Chelmsford CM1 1AT, UK). Except in cases of safety, when an archdeacon can give permission on the advice of a professional tree surgeon, no tree may be felled without a faculty. The issuing of a faculty involves a number of checks, including a review and possible visit by the Diocesan Advisory Committee and an opportunity for the public to object. The final decision is made by the Consistory Court. All this is in addition to any permission required from the local authority. Although this is at a certain cost in the loss of the natural and its spontaneity, the ecclesiastical system does speak up on behalf of the voiceless trees.

These legal safeguards give substance to the maxim that possession is nine-tenths of the law. A tree possesses life and this should not be taken without good cause being duly established. In the human paradigm, 'Thou shalt not murder', a number of exceptions may be accepted, e.g. self-defence, war, judicial execution, withdrawal of medical treatment. When life may be taken can be viewed as an issue of proportionality. The benefit gained, or evil remedied, should be of greater value than the life lost, e.g. the life of the innoocent is valued more than the life of the criminal, the peaceful death more than the poor quality of continued life. In the case of a tree on this analogy, its life is arguably less valuable than that of a human, and so it should be felled if it threatens human life, even if only at the level of a non-negligible risk. And this is evidenced by the lower safeguards operated if health and safety is invoked as the reason for felling a tree. Conservationists may attempt to argue that if the tree is well away from where people are likely to go, perhaps in a distant section of a churchyard, the wildlife benefit of the tree, even if dead, outweighs the minimal risk to people. The tree could at least merely be topped rather than felled.

But can this analogy of a tree's possession of life be pressed further than is usually accepted? Where trees threaten the fabric of the church building, permission to fell is usually granted. The popular presumption, largely unquestioned, is in favour of the building because of the costliness of building work by comparison to the cost of a new tree, amongst other reasons. In this argument, the life of the tree is no longer its own possession but belongs to the site custodians. But there is some recognition of the validity of the analogy of the tree's own possession of life, because in the case of a boundary wall it is often expected that the wall should be altered to accommodate the tree. Here the loss of the tree is accepted as disproportionate to the repair costs to the wall.

At the level of botanical diversity, an area with a tree is probably less species rich than an area with open grassland. However, the loss of life of the tree seems disproportionate to the mere potential of new life in the herb layer. Yet, the removal of saplings seems acceptable and has been encouraged. The saplings do not seem to share the same status as a mature tree and the grass has not yet been adversely affected by shade. It seems better not to let a new life begin than to take it away once it has established, so new tree planting has been gently discouraged in churchyards.

Tree protection is an encouraging example of how society can make moral decisions in favour of nature conservation. The question remains why other plants and animals, or habitats, do not receive similar protection under the law.

Whether to repoint a wall

Some of the most intense cases of conflict, which have tended to be resolved only on the basis of power, have been between conservationists of two sorts: nature conservationists and cultural conservationists. In fact, in many of the circles I move in, the word 'conservation' is applied automatically to buildings and works of art and not to nature.

Among battles lost I count the removal of the moss *Racomitrium heterostichum* from a church's roof tiles. This moss was only known in Essex from this one site, although it is common in the west of the British Isles. It was claimed that the moss was encouraging the foliation of the tiles. Another church eliminated its population of Wall Rue, *Adiantum ruta-muraria*, which is uncommon in Essex, when it repointed the stonework. Another church has tried to care for a liverwort, *Scapania undulata*, that had been thought to be extinct in Essex. A population of this species grows in a brick gulley at the foot of the church wall, and these bricks have been temporarily removed to protect the liverwort while the wall is being repointed. Unfortunately, in addition to the repointing, the church is installing gutters so that the liverwort, once returned, will no longer have the benefit of run-off from the roof – a source of water that presumably had made it possible for it to survive in our dry county. Its days are now numbered. In all these cases the maintenance of the building was claimed to take precedence.

Some wildlife conservationists have appealed to the model of cherishing works of art for a motive to cherish nature (e.g. Lawton, 1991; and Hargrove, 1989, among philosophers), and there is growing interest in the aesthetics of environmental conservation (e.g. Kemal and Gaskell, 1993; Haldane, 1994; Lee, 1995). But the conservation of works of art can come into conflict with the conservation of our natural heritage. People may claim that each work of art, at least if it is a church, is unique while individual organisms are just representatives of a species (though most would hestitate to apply this argument to humans). There is also the pragmatic argument, that if the church is to continue as a site for wildlife at all, the building and its congregation must be maintained. This argument has been made explicitly by bat conservationists to justify some interference to bat roosts in churches (Sargent, 1995). If some wildlife has to be lost, it is for the good of the rest.

Generally, the strength of people's commitment to a church is not based specifically on its uniqueness as an artefact but on the power of its story as a work of art. This is consistent with Danto's view, that artworks are objects that *say* something about the world (Danto, 1981). Church buildings say a great deal, about divine transcedence and immanence, about the identity and history of a local community, amongst other things. Danto believes that artworks require interpretation to communicate their stories and churches have many interpreters, including often a local guidebook. These interpreters bring alive the stories the building tells.

Our churchyard conservation group has discovered that if the wildlife is to be accorded similar status it must have its own story to tell. There are a plethora of interpreters of nature around, but most people still seem unable to read the natural world around them, in their own locality. People do not recognize the different species and how they dwell with one another. They are not aware of the local distinctiveness of the plant and animal communities or their history of change and continuity. Many believers have yet to consider the subtlety of the story told of God the Creator by the creation itself. If the guidebook or noticeboard explains the wildlife of the churchyard, visitors will understand why patches of long grass have been left, for example. When a naturalist visits a church and shows people the flowers in the grass or catches a bat, then eyes are opened. This experience of wonder elicits care and even the bat droppings may then be accepted.

Once the significance of the natural heritage has been seen, the vision caught, then it is possible to consider the claims of wildlife alongside the claims of the cultural heritage. Only then is it psychologically effective to refer to first principles, secular or theological. If these are presented earlier, they are often experienced as a claim to the moral high ground and so resistance is generated. Jonsen and Toulmin (1988) consider that the popularity of arguing from first principles is due to its rhetorical support of the rigorist cause. But presenting principles at the outset may be unnecessary as the relationship between first principles and case-by-case decisions is not deductive but somewhat reflexive. In practice, our initial stances are found wanting when applied to a case, and our general principles develop as they are educated by experience and interaction with other opinions. For many people their principles may often be codifications and justifications for convictions that have developed as sensitivity to the natural world has grown.

The literature on environmental ethics from a secular perspective is enormous. In this publication the contributions from Holland (1995) and Comstock (1995) in particular give an entry to some of the debates. The value of nature or its parts or processes may be claimed to be instrumental, i.e. nature benefits humans, either in a material way sustaining human life and wealth or through its scientific value, its beauty or its power to uplift the human spirit. The value of nature may also be intrinsic, perhaps on grounds of sentience, the will to live, or just its independent existence. There are many anthologies of papers, one of which being Pierce and VanDeVeer (1995). Among the justifications for environmental care that can be used among church people is an appeal to stewardship (as exemplified in this publication by DeWitt, 1995; Waters, 1995; see also Hall, 1986; Berry, 1991, 1995; Wilkinson, 1991; and Osborn, 1993). But there are many other Christian arguments for environmental care (contra Norton, 1991). One has been to stress the sacramental presence of God in creation and humankind's priestly role (e.g. McDonagh, 1986; Gregorios, 1987; Clark 1993). Going further, some have argued that the world is God's 'body' (Jansen, 1984; McFague, 1993), or have used process philosophy (Birch and Cobb, 1981; Haught, 1993), or feminism (Primavesi, 1991; Ruether, 1993), or New Age thought (Fox, 1991). Other important works on environmental theology include Moltmann, 1985; Santmire, 1985; Nash, 1991; and Murray, 1992.

When the natural and the cultural stories are both told, it also becomes possible to consider compromises. In some instances, a compromise can be a recognition of mutual dependency. Then it is not a second best, accepted as a political necessity dictated by a balance of power. A wall may need to be repointed, but it can be done in a way that may assist the wildlife to continue. The work could be done in phases, giving a chance for the plants and animals that live on it to cross back and colonise the new work, decorating it by their presence. If bats use the wall for winter hibernation, the work should be done in the summer; while if the wall contains the nests of hymenopterans, it should be repointed after the summer when most of the nests will have been abandoned. Repair need not be too vigilant, it probably never was in the past, and some partially crumbling wall sections could be left available. Even death-watch beetles, *Xestobium rufovillosum*, can be left a home in the dead timber around the churchyard when the church timbers are treated. A compromise should not inevitably be thought of as a fall from the ideal, but as a way of allowing the mutual survival of several interests. In the end, the wildlife does need the building and the people who use it – and they need the wildlife, at the very least to give meaning to the church's setting and the people's lives.

Communal ethics

These case-studies have shown something of the casuistical style of debate. The social mechanisms involved in these cases are complex. Often we absorb our principles as rather shadowy attitudes shared with the social groups we identify with and the debates mirror our social divisions, e.g. amateur naturalists versus professional ecologists, ecologists versus art historians. Each group tends to interpret churchyards using its own categories. This paper has been dominated by terms such as grassland and habitat and species. A historian might speak of Roman villa sites, ancient religious sites or recent enclosures. These various social groupings, when they are articulate, tend to structure the debate so that the case of churchyards becomes assimilated to the well-rehearsed arguments. Those who see churchyards as still vital religious sites tend to be less articulate. Popular culture may attempt to quietly and stubbornly resist turning churchyards into nature reserves or historic monuments, but it is unlikely to maintain its resistance in the long term. The institutional church is largely silent on the issue as, regretably, it is retreating from engagement with other sections of society. As a result, there is a vacuum where there should be a distinctive concept of the nature and purpose of a churchyard.

The way forward, I believe, lies with engagement, and not just for the church. In a highly pluralist society such as our own it is tempting to restrict debate to the common ground, to management ideology and what can be determined by science. But if MacIntyre's argument (1981) that I presented earlier is right, this is actually a power contest dressed up as a debate. There must be acknowledgement of the ethical dimension in decision making if the decisions are to be moral ones. Ethics is a communal activity, different people and groups, each with different insights, debating and learning together. Those who wish to advocate the conservation of nature, or who wish to refine their judgements of what the conservation of nature might be, have every right and need to be part of the communal debate. We must put our cases, tell our stories. We must also listen and learn from what others have to tell. Sometimes these debates are obviously moral ones, sometimes they are conducted in coded language. Bringing out the hidden moral dimensions of our conversations will help the process. For ethics does not stand still. Odysseus once hung his slave-girls at his pleasure, as Aldo Leopold (1949) reminded us as evidence that our sensibilities do develop.

The awareness of the moral claims of the natural world is developing in our own time and it is no longer so acceptable to damage wildlife at our pleasure as it was. The next challenge is to give nature a voice in the debate, or, rather, to pay attention to nature's own voice. Our churchyard group has not made much progress in doing this so far. We have to learn to watch for changes in the natural world and to try and catch their meaning. By this I mean, we ought to institute much more careful monitoring of the creatures and their habitats in our churchyards. We should then attempt to discern the significance of these changes, what are their causes, what they presage. Often the changes in nature are a sort of conversation with our human impact. Sadly, the changes are often analogous to a cry of pain. But it need not be, and I have the joy of visiting churchyards where sensitive decisions on the part of their custodians have led to a great flourishing of wildlife.

Acknowledgements

The author is very grateful for the comments of David Bain, Eve Dennis, Clare Palmer, George Peterken, an anonymous referee, and his fellow editor (as well as for much else).

References

Advisory Council for the Churches Ministry (1974) *Teaching Christian Ethics; an Approach.* London: SCM Press.

Barker, G.M.A. (1972) *Wildlife Conservation in the Care of Churches and Churchyards.* London: CIO Publishing.

Berry, R.J. (1991) Christianity and the environment (with bibliography on environmental issues). *Science and Christian Belief* **3**, 3–18.

Berry, R.J. (1995) Creation and the environment. *Science and Christian Belief* **7**, 21–43.

Birch, C. and Cobb, J.R. (1981) *Liberating Life: from The Cell to The Community.* Cambridge: Cambridge University Press.

Botkin, D.B. (1990) *Discordant Harmonies: a New Ecology for The Twenty-First Century.* New York: Oxford University Press.

Burman, P. and Stapleton, H. (1988) *The Churchyards Handbook, 3rd edn.* London: Church House Publishing.

Callicott, J.B. (1990) The case against moral pluralism. *Environ. Ethics* **12**, 99–124.

Clark, S.R.L. (1993) *How to Think About The Earth: Philosophical and Theological Models for Ecology.* London: Mowbray.

Comstock, G.L. (1995) An extensionist environmental ethic. *Biodiv. Conserv.* **4**, 827–37.

Cooper, N.S. (1989) Wildlife conservation in churchyards. *Crucible* **28**, 116–18.

Cooper, N.S. (1995) *Wildlife in Church and Churchyard.* London: Church House Publishing.

Danto, A. (1981) *The Transfiguration of the Commonplace.* Cambridge, MA: Harvard University Press.

Dewar, L. (1968) *An Outline of Anglican Moral Theology.* London: Mowbrays.

Dennis, E. (1993) The living churchyard – sanctuaries for wildlife. *British Wildlife* **4**, 230–41.

DeWitt, C.B. (1995) Ecology and ethics: relation of religious belief to ecological practice in the Biblical tradition. *Biodiv. Conserv.* **4**, 838–48.

Ehrenfeld, D. (1993) *Beginning Again.* Oxford: Oxford University Press.

Fox, M. (1991) *Creation Spirituality: Liberating Gifts for The Peoples of The Earth.* San Francisco: Harper San Francisco.

Gregorios, P.M. (1987) *The Human Presence: Ecological Spirituality and the Age of the Spirit.* New York: Amity House.

Grove-White, R. and Szerszynski, B. (1992) Getting behind environmental ethics. *Environ. Values* **1**, 285–96.

Haldane, J. (1994) Admiring the high mountains. *Environ. Values* **3**, 97–106.

Hall, D.J. (1986) *Imaging God: Dominion as Stewardship.* Grand Rapids MI: Eerdmans.

Hambler, C. and Speight, M.R. (1995) Biodiversity conservation in Britain: science replacing tradition. *British Wildlife* **6**, 137–47.

Hargrove, E.C. (1989) *Foundations of Environmental Ethics.* Englewood Cliffs: Prentice Hall.

Haught, J.F. (1993) *The Promise of Nature: Ecology and Cosmic Purpose.* New York: Paulist Press.

Holland, A. (1995) The use and abuse of ecological concepts in environmental ethics. *Biodiv. Conserv.* **4**, 812–26.

Howarth, J.M. (1995) Ecology: modern hero or post-modern villain? From scientific trees to phenomenological wood. *Biodiv. Conserv.* **4**, 786–97.

Idle, E.T. (1995) Conflicting priorities in site management in England. *Biodiv. Conserv.* **4**, 929–37.

Jansen, G. (1984) *God's World, God's Body.* London: Darton Longman & Todd.
Jonsen, A.R. and Toulmin, S. (1988) *The Abuse of Casuistry: a History of Moral Reasoning.* Berkeley: University of California Press.
Kemal, S. and Gaskell, I. (1993) Nature, fine arts, and aesthetics. In *Landscape, Natural Beauty and the Arts.* (S. Kemal and I. Gaskell, eds) pp. 1–42. Cambridge: Cambridge University Press.
Kirk, K.E. (1936) *Conscience and its Problems, 2nd edn.* London: Longmans Green & Co.
Lawton, J.H. (1991) Are species useful? *Oikos* **62**, 3–4.
Lee, K. (1995) Beauty for ever. *Environ. Values* **4**, 213–26.
Leopold, A. (1949) *A Sand County Almanac and Sketches Here and There.* New York: Oxford University Press.
McDonagh, S. (1986) *To Care for The Earth.* London: Geoffrey Chapman.
McFague, S. (1993) *The Body of God: an Ecological Theology.* London: SCM Press.
MacIntyre, A. (1981) *After Virtue: a Study in Moral Theory.* London: Duckworth.
Maunder, M. (1992) Plant reintroduction: an overview. *Biodiv. Conserv.* **1**, 51–61.
Midgley, M. (1992) Beasts versus the biosphere? *Environ. Values* **1**, 113–21.
Moltmann, J. (1985) *God in Creation: an Ecological Doctrine of Creation.* London: SCM Press.
Murray, R. (1992) *The Cosmic Covenant.* London: Sheed and Ward.
Nash, J.A. (1991) *Loving Nature: Ecological Integrity and Christian Responsibility.* Nashville: Abingdon Press.
Norton, B.G. (1991) *Toward Unity Among Environmentalists.* Oxford: Oxford University Press.
Osborn, L.H. (1993) *Guardians of Creation: Nature in Theology and the Christian Life.* Leicester: Apollos.
Palmer, C. (1992) Stewardship: a case study in environmental ethics. In *The earth beneath: a critique of green theology.* (I. Ball, M. Goodall, C. Palmer and J. Reader, eds). London: SPCK.
Pierce, C. and VanDeVeer, D., eds (1995) *People, Penguins, and Plastic Trees,* 2nd edn. Belmont CA: Wadsworth Publishing.
Peters, R.H. (1991) *A Critique for Ecology.* Cambridge: Cambridge University Press.
Primavesi, A. (1991) *From Apocalypse to Genesis: Ecology, Feminism and Christianity.* Tunbridge Wells: Burns and Oates.
Rackham, O. (1986) *The History of the Countryside.* London: J.M. Dent and Sons.
Ruether, R.R. (1993) *Gaia and God: an Ecofeminist Theology of Earth Healing.* London: SCM Press.
Ruse, M. (1994) From belief to unbelief – and halfway back. *Zygon* **29**, 25–35.
Santmire, H.P. (1985) *The Travail of Nature: the Ambiguous Ecological Promise of Christian Theology.* Philadelphia: Fortress Press.
Sargent, G. (1995) *The Bats in Churches Project.* London: The Bat Conservation Trust.
Schrader-Frechette, K. and McCoy, E.D. (1993) *Method in Ecology, Strategies for Conservation.* Cambridge: Cambridge University Press.
Stone, C. (1974) *Should Trees Have Standing?* Los Altos, CA: William Kaufman.
Stone, C. (1988) *Earth and Other Ethics: the Case for Moral Pluralism.* New York: Harper and Row.
Waters, B. (1995) Christian theological resources for environmental ethics. *Biodiv. Conserv.* **4**, 849–56.
Weiner, J. (1995) On the practice of ecology. *J. Ecol.* **83**, 153–8.
Wilkinson, L., ed. (1991) *Earthkeeping in the '90s: Stewardship of Creation.* Grand Rapids, MI: Eerdmans.

13

Conflicting priorities in site management in England

E.T. IDLE

English Nature, Northminster House, Peterborough PE1 1UA, UK*

There are two main designations of 'protected' areas in the UK – nature reserves, of various kinds, and special nature conservation areas, known as Sites of Special Scientific Interest. General approaches to choices of priorities and the resolution of conflict in the management of these areas are described and difficulties identified. Similar problems arise when considering the wider role of 'protected' areas in national nature conservation policies, e.g. biodiversity targets and European Habitats and Species Directive objectives. Because choices and priorities stem from underlying values, the public must be involved in the identification of objectives for both 'protected' areas and the rest of the land surface. The Natural Areas programme being developed by English Nature provides a mechanism whereby people may be involved in characterising the wildlife of 'their' area and in identifying targets for its maintenance and enhancement. The use of land and the management practices associated with it are the major factors influencing the nature conservation value of 'protected' areas and their wider context.

Keywords: protected areas; management choices; nature conservation; policy objectives; Natural Areas; integrated land use

Introduction

England is a relatively small but highly populated country with a history of industrial and commercial development and intensification of land use, particularly during the 20th century. No areas of land or water which could be described as 'wilderness' or 'natural', in the sense that they reflect very limited human interference, are left. Indeed transformation from the original state of nature began several hundreds, if not thousands, of years ago. The growing intensity of land use in the last 50 years has left 5–10% of the land surface in what can be described as a semi-natural state in contrast to the majority of the land dominated by the intensive cultivation of agricultural crops. This has resulted in well-documented reductions in biodiversity in both communities and populations of plant and animal species (Brown, 1992).

In Great Britain the 1949 National Parks and Access to the Countryside Act gave legal force to the need for a national series of nature reserves. These were to be areas of land or water which fulfilled two broad purposes. First, to provide special opportunities for study and research; and second, to preserve flora, fauna, geology and physiography. Nature reserves had a different purpose from National Parks, but in practice the two designations

* English Nature (the Nature Conservancy Council for England) is the statutory advisor to the government on nature conservation in England and promotes the conservation of England's wildlife and natural features.

were seen as potentially mutually supportive. From the start the need for management within nature reserves was accepted and the dangers of the loss of interest and value through 'lack of knowledge, indifferent management and stupidity' was understood (Cmd. 7122, 1947). Nature reserves, which had lost their original natural state, required management in order to retain the value of their semi-natural state, within their relatively confined boundaries.

Background

In the UK there are now two main groups of sites which are managed primarily for nature conservation – nature reserves including National Nature Reserves, voluntary organization reserves and local authority reserves – and special sites, called Sites of Special Scientific Interest (SSSIs). Protection, management and enhancement of these areas lie at the heart of nature conservation strategy for England. They are integral to the achievement of nature conservation goals for the foreseeable future. Many countries have similar networks of protected sites and areas where ownership and management may be under public or Government control and safeguards are maintained against a wide range of harmful developments. Attempts have been made, particularly by IUCN, to standardize nomenclature and criteria for selection of such sites, but they have not been entirely successful because of difficulties with terminology (IUCN, 1994). This is partly because of the differing social, economic, political and geographical/historical background of differing countries and the related issues of size and scale of conservation areas, and the objectives for such sites or areas in an often unstated nature conservation policy. For example, only recently have National Parks in England and Wales been given a clear nature conservation responsibility (Environment Bill, 1994), despite the fact that in 1947 National Parks and nature reserves were seen as complementary instruments to promote national nature conservation policies and objectives. In contrast, in much of the rest of the world National Parks have nature conservation as one of their primary objectives. However despite their large size, their role in meeting wider nature conservation objectives is not clear. As pressures resulting from changes in their immediate environment grow, the extent to which National Parks can be left to develop on their own is questioned. In nature reserves, National Parks and protected areas throughout the world, management appears increasingly necessary. Where management is required, choices have to be made about objectives and methods of achieving them.

In England, there are now 157 National Nature Reserves. Management of them is controlled, though not always carried out, by English Nature (Table 1). English Nature owns about 20% of the total area of National Nature Reserves (Table 2). Altogether National Nature Reserves occupy less than 0·5% of the land surface of England. Some National Nature Reserves are held in a mixture of all three forms of tenure: ownership, lease and agreement.

In a review of the most important nature conservation sites in Great Britain, Ratcliffe (1977) outlined the criteria to be used in evaluation of site importance. The purpose of the review was to 'translate in practice, criteria to guide the selection of key sites' for nature conservation. It accepted that the definition of 'best' sites is difficult in terms of measurable, specific qualities and involves many-sided value judgements in which standards are essentially relative and not absolute. The criteria used in site assessment were size, diversity, naturalness, rarity, fragility, typicalness, recorded history, ecological

Table 1. National Nature Reserves in England

Total number of NNRs	157	(at 30 September 1994)
Total hectarage of NNRs	61 558	(0·45% of England's land surface)
Total staff employed on NNRs		
Site managers	62	
Estate workers	40	

Source: *English Nature Facts and Figures* (Autumn 1994).

position, potential value and intrinsic appeal. To a greater or lesser degree these criteria for nature conservation value have continued to be used to the present day. Some, such as size, diversity and rarity, have proved much easier to use in evaluation procedures, while others, typicalness and intrinsic appeal, have been difficult (Usher, 1986). However, as Ratcliffe stated, all carry a degree of subjectivity but can be applied systematically to biological or other information about nature conservation sites.

The objectives of management of nature reserves are derived from an evaluation of their nature conservation interest using the criteria just described. The objectives are encapsulated in the reserve management plan (NCC, 1988). Frequently choices have to be made between potentially conflicting objectives, particularly on whether to interfere with habitats and species and manage them – or not. Decisions have to be made on whether the rare and varied species of a site are more desirable than the processes of ecological change which will occur within the site. Ecological change may threaten the continuation of the rare. Furthermore, decisions are often required on whether to favour one species, group of species or habitat and not another. Thus on a relatively small nature reserve where the presence of a rare species depends on the maintenance of a particular, often early, stage of vegetational succession, management will be necessary to maintain that species. However, the implementation of appropriate management regimes to safeguard the species will mean that any wish to allow natural vegetation development will be lost. Conversely, to allow natural development or succession will set at risk the future of the rare species in that site. In these circumstances, ways need to be found of balancing conflicting values. Often the way out has been to divide the site or nature reserve and pursue one value in one part and another in the remainder. With management plans the choice is set out in a rationale

Table 2. Ownership of National Nature Reserves in England

Tenure	Hectares
Owned by English Nature	14 439
Leased by English Nature	29 808
Nature Reserve Agreements with English Nature	13 702
Held by an approved body (Section 35: Wildlife & Countryside Act 1981)	3609
	51 558
Total number of NNRs	157

Source: *English Nature Facts and Figures* (Autumn 1994).

Table 3. National Nature Reserve management regimes in England

Management regimes	Number of reserves using this regime
Coppicing	63
Traditional woodland management	59
Ride/path/glade maintenance	93
Scrub control	104
Controlled grazing	94
Controlled burning	24
Cutting/mowing/haymaking	93
Water level control	54
Ditch/dyke maintenance	51
Water edge management	61
Planting/stabilization	42
Non intervention	106
Total number of National Nature Reserves (28 February 1995) =	157

Source: Annual Report of National Nature Reserves and Marine Nature Reserves, available from English Nature.

which follows an evaluation of nature conservation interest in comparison with other similar sites. Because in England, at present, nature reserves are small and represent fragments of formerly widespread vegetation, management is usually necessary if nature conservation value is to be maintained. The main management regimes required to implement the choice of objectives can be summarized (Table 3).

It is clear from this Table that some nature reserves have more than one management regime. In some cases there may be several, reflecting detailed management programmes. On privately-owned land which is not designated for nature conservation purposes such detailed management regimes are often not possible. Only on areas specifically managed for nature conservation can these regimes be carried out. Private landowners are often not able or cannot afford such management. Yet the summary provides a national framework of nature reserve management and is particularly useful for information exchange, organizational learning and the dissemination of management techniques to farmers and foresters and the main owners of sites of special wildlife interest.

Sites of Special Scientific Interest are areas of land and water which are identified as 'special' because of their fauna, flora, geology and landforms. Many are privately owned. All are subject to formal consultation on planning development and potential management change. They form a network of over 3800 sites across the country and are an effective way of safeguarding important areas from planning development. However, ensuring appropriate management of them to maintain their nature conservation interest can be more problematic. Together with nature reserves, SSSIs represent the most important wildlife sites in England.

Figure 1 illustrates the relationships between National Nature Reserves, SSSIs and other land. The total area of land 'protected' within these areas amounts to about 7% of England. This does not include land within National Parks and does not include areas where conservation conditions are applied to agricultural and forestry practices. If these

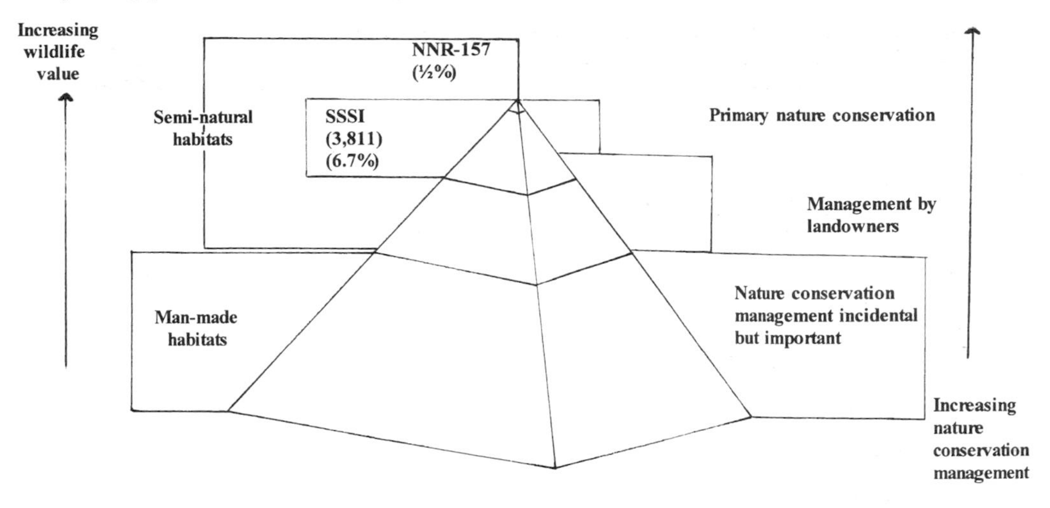

Figure 1. Site designation and management in England (from Nature Conservancy Council, 1984).

areas are added, the total extent of designated or 'protected' areas amounts to 10–15%. Comparisons of the extent of protection with other countries is difficult. Like cannot easily be compared with like because of the problem of defining terms. Furthermore, protected areas are only one of a series of instruments or mechanisms which deliver nature conservation policies or objectives. Others include species licensing and controls and conservation conditions allied to financial support to agriculture.

Recent thinking and development

English Nature's overall objective is 'to create a situation where the characteristic biological diversity and natural features of England are maintained and enriched, across their traditional range' (English Nature, 1993).

This of course begs the question of what we mean by 'characteristic biological diversity' and 'traditional range'. In England geographical and ecological variation across the country is well marked. The moss and lichen-rich oak woodlands in the west of the country contrast sharply with the mixed deciduous woodlands overlying deeper soils in the east. Equally the flora and fauna of the chalk downlands of the south differ from the tall hay meadows of the northern valleys. The biological diversity of these areas can be described and their traditional range demonstrated, taking account of the changes that have taken place over recent years.

English Nature is in the process of developing a 'Natural Areas Programme' based on a Natural Areas map (see Fig. 2). This builds on the land use/land cover work of Stamp (1946) and 19th and early 20th century botanists such as Watson (1883) and Druce (1932). Natural Areas reflect the geological foundation, natural systems and processes and the wildlife of different parts of England. Within them they have characteristic patterns of land use and have an identity to which local people can relate. They provide a basis for the description of the flora and fauna and importantly provide a mechanism of identifying targets and goals for nature conservation which have local support and involvement. The problem then is to identify the contribution that nature reserves and other conservation mechanisms make to the achievement of these goals.

English Nature has defined more clearly the roles expected from important nature conservation sites. They are:

(i) To reduce the risks of further losses of nature conservation interest.
(ii) To provide a core resource from which to build the wider nature conservation character of a locality.
(iii) To provide opportunities for experiencing nature and participating in its management.
(iv) To provide oppportunities for demonstration, training and study of nature conservation management.
(v) To act as reference points and controls in wider studies and for monitoring.

These functions or policy objectives apply to both nature reserves and special sites. Indeed they probably apply to all nature conservation or 'protected' areas.

This prompts two other questions. Firstly, Does it matter who owns these areas? If private owners are prepared to manage their land in ways that produce desired nature conservation results, where and how should conservation organizations be involved? Often a degree of appropriate management by private owners is possible, particularly if financial support in one form or another is available from Government. However, many private owners of land have a range of primary purposes which pose some risk to the sustainability of the wildlife. Where it is not possible to get the right management of important sites there is a *prima facie* case for ownership or management involvement by a conservation organization.

The second question is, Who decides what the objectives for natural areas should be and the contribution nature reserves and special sites make to them? As was expressed several times at the Ecologists and Ethical Judgements Symposium at INTECOL 1994 people must be involved, and not just in legitimising objectives which ecologists and conservationists have decided.

English Nature, and its predecessor the Nature Conservancy Council, has involved people in communities in its work for many years, particularly on National Nature Reserves. This has mostly involved volunteers in wildlife survey work, management of habitats and contacts with visitors and the public. Only rarely have people been involved in the definition of objectives for nature reserves and the management programmes needed to support them. In some cases where management has resulted in species culling programmes, local communities have expressed serious concern and opposition. These concerns have been persistent and vociferous and mostly focus on the idea that on nature reserves animals should not be killed, or even in some cases controlled. Finding ways of even recognizing underlying differences in the approach to these problems and subsequently resolving them is a major challenge for English Nature. It may be that recent work on philosophical aspects of conservation will be of assistance in this problem.

The future

The clarification of the function and role of important wildlife sites paves the way for considering the contribution each site makes to the Natural Area of which it is a part. In 1995 English Nature expects to embark on this process and to review National Nature

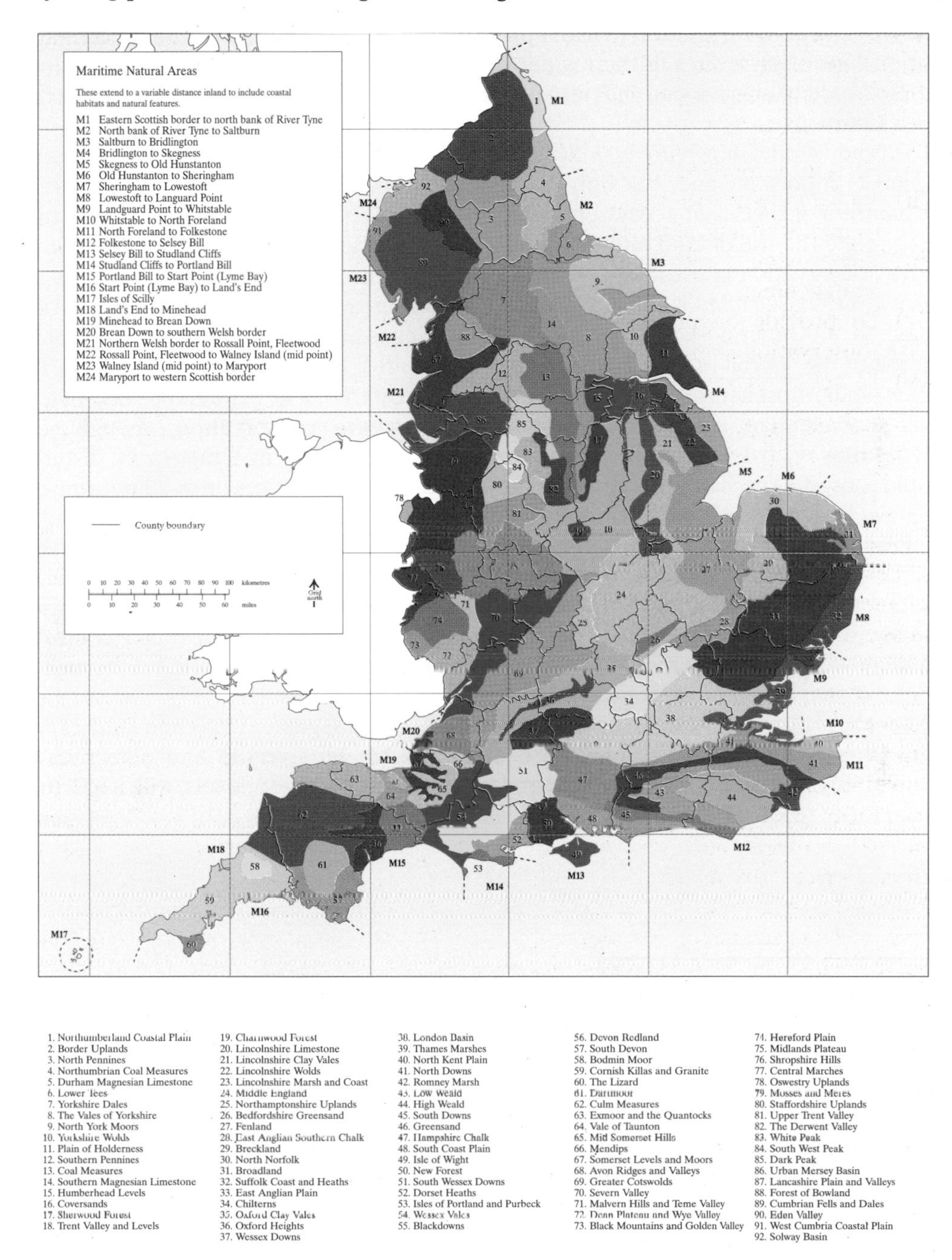

Figure 2. Natural areas.
The grid on this map is the National Grid taken from the Ordnance Survey map with the permission of The Controller of Her Majesty's Stationery Office.

Reserves and SSSIs in relation to wider policy objectives for wildlife. At the same time the relationships of these sites to their wider spatial and policy contexts are being developed and used to help achieve national targets for biodiversity and the European Habitats and Species Directive.

The crude model shown in Fig. 3 illustrates these relationships. Within human timescales the most important, and certainly most influenceable, factors which determine the combination of species or ecosystems which produce nature conservation value are those which relate the use of land and the management practices employed in that use. So the nature conservation interest, nc, of each of these areas is summarised as $nc = f(u, m)$, where u is the use and m is the management of each area. Nature reserves or special sites equate to the innermost box, where the nature conservation value, nc is maintained, because of the special use and management of that area, u, m. In England, particularly in the lowlands, most nature reserves or special sites are part of a larger unit box nc_2. In these areas management is often unsympathetic to nature conservation, though some contribution is made, e.g. in providing feeding areas for birds or temporarily disturbed ground for some species of insects and plants. Management practices on neighbouring land can have profound influences on the nature conservation value of nature reserves and special sites. Thus although the nature conservation value of land surrounding reserves is less, it can be defined by a similar equation $nc_2 = f(u_2, m_2)$. The interest of the special site and its context is a combination of nc, and nc_2, i.e. it depends on the management and use of both site and context. The outer box, nc_3, represents the landscape or policy context of individual farm or management units. Here again the contribution to nature is through the use and management practices carried out.

At each level, nature reserve or special site, farm or management unit and landscape or policy scale, conflicts in priorities arise. If targets for biodiversity and objectives for sustainability are to be met conflicts and priorities at all three levels will need to be satisfactorily resolved.

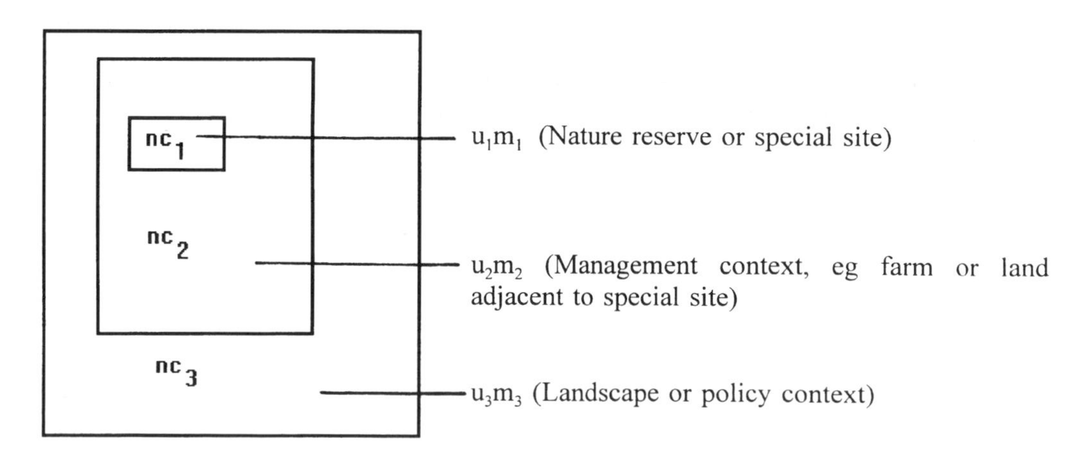

Figure 3. 'Protected' sites and their contexts. The nature conservation value nc is a function of the uses u and management m of all three boxes.

Conclusions

Conflicts, priorities and choices of objectives for nature conservation are clear at the level of individual nature reserves or other 'protected' areas. In intensively managed and often fragmented landscapes it is impossible to isolate the reserve and its objectives from its immediate environment and similar conflicts, priorities and choices often have to be faced. The same applies to landscape or policy considerations. In the past, and possibly even today in some countries, attempts have been made to resolve these problems by strictly separating the conservation of nature from the main land uses of agriculture, forestry, housing, roads and industry. Where protected areas are large, such as in some National Parks, such solutions may seem to work at first. However, the choices which have to be made by conservation managers are not always acceptable to the public. Furthermore many people regard 'nature' as part of their quality of life. They may accept that in some cases it may need to be confined in parks or nature reserves, but many people prefer to be able to enjoy the benefits of wildlife without having to make sometimes lengthy journeys to witness it. This does not mean that nature reserves and 'protected' areas are unimportant, but it does mean that for everyone to be able to enjoy nature, for whatever reasons, the priorities, conflicts and choices which occur on nature reserves must be faced at every level.

References

Brown, A. (1992) *The UK Environment.* London: HMSO.

Cmd. 7122 (1947) *Conservation of Nature in England and Wales. Report of the Wild Life Conservation Special Committee (England and Wales).* London: HMSO.

Druce, G.C. (1932) *The Comital Flora of The British Isles.* Arbroath.

English Nature (1993) *English Nature's strategy for the 1990s.* Peterborough: English Nature.

Environment Bill. Introduced by House of Lords in 1994. (HoL Bill 30, 51/3).

IUCN (1994) *Guidelines For Protected Area Management Categories.* Cambridge: NPPA/WCMC.

Nature Conservancy Council (1984) *Nature Conservation in Great Britain.* Peterborough: Nature Conservancy Council.

NCC (1988) *Site Management Plans For Nature Conservation: a Working Guide.* Peterborough: Nature Conservancy Council.

Ratcliffe, D.A. (1977) *A Nature Conservation Review, Volumes 1 and 2.* Cambridge: Cambridge University Press.

Stamp, L.D. (1946) *Britain's Structure and Scenery.* London: Collins.

Usher, M.B. (1986) *Wildlife Conservation Evaluation.* London: Chapman and Hall.

Watson, H.C. (1883) *Topographical Botany.* London: Quaritch.

Index

Page references in **bold** refer to figures; those in *italics* refer to tables.